VIBRATION FUNDAMENTALS

PLANT ENGINEERING MAINTENANCE SERIES

Vibration Fundamentals
R. Keith Mobley

Root Cause Failure Analysis
R. Keith Mobley

Maintenance Fundamentals
R. Keith Mobley

VIBRATION FUNDAMENTALS

R. Keith Mobley

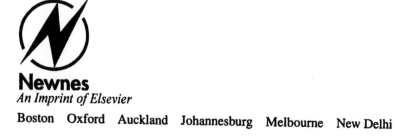

Newnes
An Imprint of Elsevier

Boston Oxford Auckland Johannesburg Melbourne New Delhi

Newnes is an imprint of Elsevier.

Copyright ©1999 by Butterworth–Heinemann

 Recognizing the importance of preserving what has been written, Butterworth–Heinemann prints its books on acid-free paper whenever possible.

Library of Congress Cataloging-in-Publication Data
Mobley, R. Keith, 1943-
Vibration fundamentals / by R. Keith Mobley.
    p. cm. – (Plant engineering maintenance series)
Includes index.
ISBN 0-7506-7150-5 (alk. paper)
1. Plant maintenance. 2. Vibration. 3. Machinery—Vibration.
I. Title. II. Series.
TS192.M626 1999
621.8'16—dc21   98-32098
CIP

British Library Cataloguing-in-Publication Data
A catalogue record for this book is available from the British Library.
The publisher offers special discounts on bulk orders of this book.
For information, please contact:
Manager of Special Sales
Butterworth–Heinemann
225 Wildwood Avenue
Woburn, MA 01801–2041
Tel: 781-904-2500
Fax: 781-904-2620
For information on all Butterworth–Heinemann publications available,
contact our World Wide Web home page at: http://www.bh.com

Transferred to Digital Printing 2007

CONTENTS

vi   Contents

Part I

THEORY: INTRODUCTION TO VIBRATION ANALYSIS

Part I is an introduction to vibration analysis that covers basic vibration theory. All mechanical equipment in motion generates a vibration profile, or signature, that reflects its operating condition. This is true regardless of speed or whether the mode of operation is rotation, reciprocation, or linear motion. Vibration analysis is applicable to all mechanical equipment, although a common—yet invalid—assumption is that it is limited to simple rotating machinery with running speeds above 600 revolutions per minute (rpm). Vibration profile analysis is a useful tool for predictive maintenance, diagnostics, and many other uses.

Chapter 1

INTRODUCTION

Several predictive maintenance techniques are used to monitor and analyze critical machines, equipment, and systems in a typical plant. These include vibration analysis, ultrasonics, thermography, tribology, process monitoring, visual inspection, and other nondestructive analysis techniques. Of these techniques, vibration analysis is the dominant predictive maintenance technique used with maintenance management programs.

Predictive maintenance has become synonymous with monitoring vibration characteristics of rotating machinery to detect budding problems and to head off catastrophic failure. However, vibration analysis does not provide the data required to analyze electrical equipment, areas of heat loss, the condition of lubricating oil, or other parameters typically evaluated in a maintenance management program. Therefore, a total plant predictive maintenance program must include several techniques, each designed to provide specific information on plant equipment.

Chapter 2

VIBRATION ANALYSIS APPLICATIONS

The use of vibration analysis is not restricted to predictive maintenance. This technique is useful for diagnostic applications as well. Vibration monitoring and analysis are the primary diagnostic tools for most mechanical systems that are used to manufacture products. When used properly, vibration data provide the means to maintain optimum operating conditions and efficiency of critical plant systems. Vibration analysis can be used to evaluate fluid flow through pipes or vessels, to detect leaks, and to perform a variety of nondestructive testing functions that improve the reliability and performance of critical plant systems.

Some of the applications that are discussed briefly in this chapter are predictive maintenance, acceptance testing, quality control, loose part detection, noise control, leak detection, aircraft engine analyzers, and machine design and engineering. Table 2.1 lists rotating, or centrifugal, and nonrotating equipment, machine-trains, and continuous processes typically monitored by vibration analysis.

Table 2.1 Equipment and Processes Typically Monitored by Vibration Analysis

Centrifugal	Reciprocating	Continuous Process
Pumps	Pumps	Continuous casters
Compressors	Compressors	Hot and cold strip lines
Blowers	Diesel engines	Annealing lines
Fans	Gasoline engines	Plating lines
Motor/generators	Cylinders	Paper machines
Ball mills	Other machines	Can manufacturing lines
Chillers		Pickle lines

continued

Table 2.1 Equipment and Processes Typically Monitored by Vibration Analysis

Centrifugal	Machine-Trains	Continuous Process
Product rolls	Boring machines	Printing
Mixers	Hobbing machines	Dyeing and finishing
Gearboxes	Machining centers	Roofing manufacturing lines
Centrifuges	Temper mills	Chemical production lines
Transmissions	Metal-working machines	Petroleum production lines
Turbines	Rolling mills, and most	Neoprene production lines
Generators	machining equipment	Polyester production lines
Rotary dryers		Nylon production lines
Electric motors		Flooring production lines
All rotating machinery		Continuous process lines

Source: Integrated Systems, Inc.

PREDICTIVE MAINTENANCE

The fact that vibration profiles can be obtained for all machinery that has rotating or moving elements allows vibration-based analysis techniques to be used for predictive maintenance. Vibration analysis is one of several predictive maintenance techniques used to monitor and analyze critical machines, equipment, and systems in a typical plant. However, as indicated before, the use of vibration analysis to monitor rotating machinery to detect budding problems and to head off catastrophic failure is the dominant predictive maintenance technique used with maintenance management programs.

ACCEPTANCE TESTING

Vibration analysis is a proven means of verifying the actual performance versus design parameters of new mechanical, process, and manufacturing equipment. Preacceptance tests performed at the factory and immediately following installation can be used to ensure that new equipment performs at optimum efficiency and expected life-cycle cost. Design problems as well as possible damage during shipment or installation can be corrected before long-term damage and/or unexpected costs occur.

QUALITY CONTROL

Production-line vibration checks are an effective method of ensuring product quality where machine tools are involved. Such checks can provide advanced warning that the surface finish on parts is nearing the rejection level. On continuous process lines such as paper machines, steel-finishing lines, or rolling mills, vibration analysis can prevent abnormal oscillation of components that result in loss of product quality.

LOOSE OR FOREIGN PARTS DETECTION

Vibration analysis is useful as a diagnostic tool for locating loose or foreign objects in process lines or vessels. This technique has been used with great success by the nuclear power industry and it offers the same benefits to non-nuclear industries.

NOISE CONTROL

Federal, state, and local regulations require serious attention be paid to noise levels within the plant. Vibration analysis can be used to isolate the source of noise generated by plant equipment as well as background noises such as those generated by fluorescent lights and other less obvious sources. The ability to isolate the source of abnormal noises permits cost-effective corrective action.

LEAK DETECTION

Leaks in process vessels and devices such as valves are a serious problem in many industries. A variation of vibration monitoring and analysis can be used to detect leakage and isolate its source. Leak-detection systems use an accelerometer attached to the exterior of a process pipe. This allows the vibration profile to be monitored in order to detect the unique frequencies generated by flow or leakage.

AIRCRAFT ENGINE ANALYZERS

Adaptations of vibration analysis techniques have been used for a variety of specialty instruments, in particular, portable and continuous aircraft engine analyzers. Vibration monitoring and analysis techniques are the basis of these analyzers, which are used for detecting excessive vibration in turboprop and jet engines. These instruments incorporate logic modules that use existing vibration data to evaluate the condition of the engine. Portable units have diagnostic capabilities that allow a mechanic to determine the source of the problem while continuous sensors alert the pilot to any deviation from optimum operating condition.

MACHINE DESIGN AND ENGINEERING

Vibration data have become a critical part of the design and engineering of new machines and process systems. Data derived from similar or existing machinery can be extrapolated to form the basis of a preliminary design. Prototype testing of new machinery and systems allows these preliminary designs to be finalized, and the vibration data from the testing adds to the design database.

Chapter 3

VIBRATION ANALYSIS OVERVIEW

Vibration theory and vibration profile, or signature, analyses are complex subjects that are the topic of many textbooks. The purpose of this chapter is to provide enough theory to allow the concept of vibration profiles and their analyses to be understood before beginning the more in-depth discussions in the later sections of this module.

THEORETICAL VIBRATION PROFILES

A vibration is a periodic motion or one that repeats itself after a certain interval of time. This time interval is referred to as the period of the vibration, T. A plot, or profile, of a vibration is shown in Figure 3.1, which shows the period, T, and the maximum displacement or amplitude, X_0. The inverse of the period, $\frac{1}{T}$, is called the frequency, f, of the vibration, which can be expressed in units of cycles per second (cps) or Hertz (Hz). A harmonic function is the simplest type of periodic motion and is shown in Figure 3.2, which is the harmonic function for the small oscillations of a simple pendulum. Such a relationship can be expressed by the equation:

$$X = X_0 \sin(\omega t)$$

where

X = Vibration displacement (thousandths of an inch, or mils)
X_0 = Maximum displacement or amplitude (mils)
ω = Circular frequency (radians per second)
t = Time (seconds).

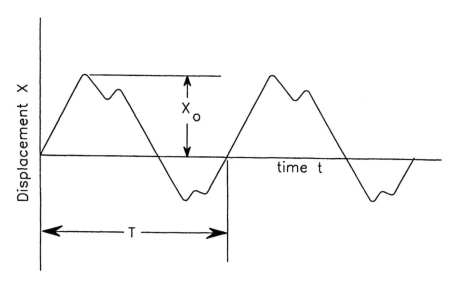

Figure 3.1 Periodic motion for bearing pedestal of a steam turbine.

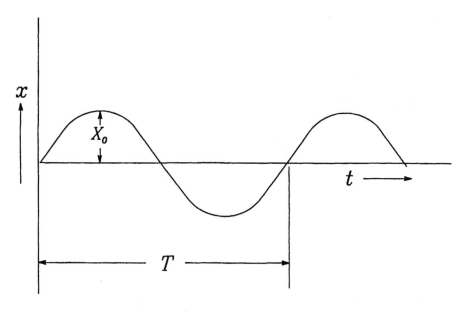

Figure 3.2 Small oscillations of a simple pendulum, harmonic function.

ACTUAL VIBRATION PROFILES

The process of vibration analysis requires the gathering of complex machine data, which must then be deciphered. As opposed to the simple theoretical vibration curves shown in Figures 3.1 and 3.2 above, the profile for a piece of equipment is extremely complex. This is true because there are usually many sources of vibration. Each source generates its own curve, but these are essentially added and displayed as a composite profile. These profiles can be displayed in two formats: time domain and frequency domain.

Time Domain

Vibration data plotted as amplitude versus time is referred to as a time-domain data profile. Some simple examples are shown in Figures 3.1 and 3.2. An example of the complexity of these type of data for an actual piece of industrial machinery is shown in Figure 3.3.

Time-domain plots must be used for all linear and reciprocating motion machinery. They are useful in the overall analysis of machine-trains to study changes in operating conditions. However, time-domain data are difficult to use. Because all of the vibration data in this type of plot are added to represent the total displacement at any given time, it is difficult to determine the contribution of any particular vibration source.

The French physicist and mathematician Jean Fourier determined that nonharmonic data functions such as the time-domain vibration profile are the mathematical sum of

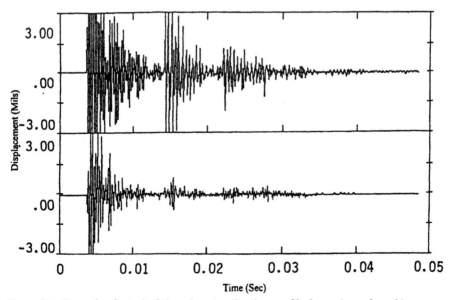

Figure 3.3 Example of a typical time-domain vibration profile for a piece of machinery.

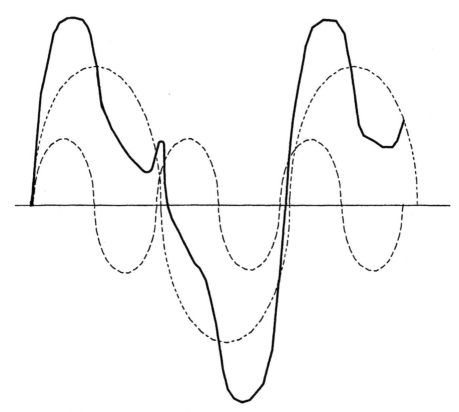

Figure 3.4 Discrete (harmonic) and total (nonharmonic) time-domain vibration curves.

simple harmonic functions. The dashed-line curves in Figure 3.4 represent discrete harmonic components of the total, or summed, nonharmonic curve represented by the solid line.

These type of data, which are routinely taken during the life of a machine, are directly comparable to historical data taken at exactly the same running speed and load. However, this is not practical because of variations in day-to-day plant operations and changes in running speed. This significantly affects the profile and makes it impossible to compare historical data.

Frequency Domain

From a practical standpoint, simple harmonic vibration functions are related to the circular frequencies of the rotating or moving components. Therefore, these frequencies are some multiple of the basic running speed of the machine-train, which is expressed in revolutions per minute (rpm) or cycles per minute (cpm). Determining

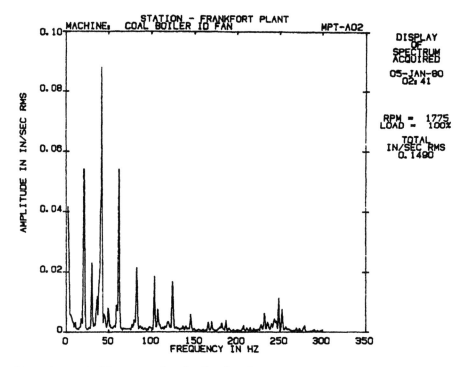

Figure 3.5 Typical frequency-domain vibration signature.

these frequencies is the first basic step in analyzing the operating condition of the machine-train.

Frequency-domain data are obtained by converting time-domain data using a mathematical technique referred to as a fast Fourier transform (FFT). FFT allows each vibration component of a complex machine-train spectrum to be shown as a discrete frequency peak. The frequency-domain amplitude can be the displacement per unit time related to a particular frequency, which is plotted as the Y-axis against frequency as the X-axis. This is opposed to the time-domain spectrum, which sums the velocities of all frequencies and plots the sum as the Y-axis against time as the X-axis. An example of a frequency-domain plot or vibration signature is shown in Figure 3.5.

Frequency-domain data are required for equipment operating at more than one running speed and all rotating applications. Because the X-axis of the spectrum is frequency normalized to the running speed, a change in running speed will not affect the plot. A vibration component that is present at one running speed will still be found in the same location on the plot for another running speed after the normalization, although the amplitude may be different.

Interpretation of Vibration Data

The key to using vibration signature analysis for predictive maintenance, diagnostic, and other applications is the ability to differentiate between normal and abnormal vibration profiles. Many vibrations are normal for a piece of rotating or moving machinery. Examples of these are normal rotations of shafts and other rotors, contact with bearings, gear-mesh, etc. However, specific problems with machinery generate abnormal, yet identifiable, vibrations. Examples of these are loose bolts, misaligned shafts, worn bearings, leaks, and incipient metal fatigue.

Predictive maintenance utilizing vibration signature analysis is based on the following facts, which form the basis for the methods used to identify and quantify the root causes of failure:

1. All common machinery problems and failure modes have distinct vibration frequency components that can be isolated and identified.
2. A frequency-domain vibration signature is generally used for the analysis because it is comprised of discrete peaks, each representing a specific vibration source.
3. There is a cause, referred to as a forcing function, for every frequency component in a machine-train's vibration signature.
4. When the signature of a machine is compared over time, it will repeat until some event changes the vibration pattern (i.e., the amplitude of each distinct vibration component will remain constant until there is a change in the operating dynamics of the machine-train).

While an increase or a decrease in amplitude may indicate degradation of the machine-train, this is not always the case. Variations in load, operating practices, and a variety of other normal changes also generate a change in the amplitude of one or more frequency components within the vibration signature. In addition, it is important to note that a lower amplitude does not necessarily indicate an improvement in the mechanical condition of the machine-train. Therefore, it is important that the source of all amplitude variations be clearly understood.

VIBRATION MEASURING EQUIPMENT

Vibration data are obtained by the following procedure: (1) Mount a transducer onto the machinery at various locations, typically machine housing and bearing caps, and (2) use a portable data-gathering device, referred to as a vibration monitor or analyzer, to connect to the transducer to obtain vibration readings.

Transducer

The transducer most commonly used to obtain vibration measurements is an accelerometer. It incorporates piezoelectric (i.e., pressure-sensitive) films to convert mechanical energy into electrical signals. The device generally incorporates a weight suspended between two piezoelectric films. The weight moves in response to vibration and squeezes the piezoelectric films, which sends an electrical signal each time the weight squeezes it.

Portable Vibration Analyzer

The portable vibration analyzer incorporates a microprocessor that allows it to convert the electrical signal mathematically to acceleration per unit time, perform a FFT, and store the data. It can be programmed to generate alarms and displays of the data. The data stored by the analyzer can be downloaded to a personal or a more powerful computer to perform more sophisticated analyses, data storage and retrieval, and report generation.

Chapter 4

VIBRATION SOURCES

All machinery with moving parts generates mechanical forces during normal operation. As the mechanical condition of the machine changes due to wear, changes in the operating environment, load variations, etc., so do these forces. Understanding machinery dynamics and how forces create unique vibration frequency components is the key to understanding vibration sources.

Vibration does not just happen. There is a physical cause, referred to as a forcing function, and each component of a vibration signature has its own forcing function. The components that make up a signature are reflected as discrete peaks in the FFT or frequency-domain plot.

The vibration profile that results from motion is the result of a force imbalance. By definition, balance occurs in moving systems when all forces generated by, and acting on, the machine are in a state of equilibrium. In real-world applications, however, there is always some level of imbalance and all machines vibrate to some extent. This chapter discusses the more common sources of vibration for rotating machinery, as well as for machinery undergoing reciprocating and/or linear motion.

ROTATING MACHINERY

A rotating machine has one or more machine elements that turn with a shaft, such as rolling-element bearings, impellers, and other rotors. In a perfectly balanced machine, all rotors turn true on their centerline and all forces are equal. However, in industrial machinery, it is common for an imbalance of these forces to occur. In addition to imbalance generated by a rotating element, vibration may be caused by instability in the media flowing through the rotating machine.

Rotor Imbalance

While mechanical imbalance generates a unique vibration profile, it is not the only form of imbalance that affects rotating elements. Mechanical imbalance is the condition where more weight is on one side of a centerline of a rotor than on the other. In many cases, rotor imbalance is the result of an imbalance between centripetal forces generated by the rotation. The source of rotor vibration also can be an imbalance between the lift generated by the rotor and gravity.

Machines with rotating elements are designed to generate vertical lift of the rotating element when operating within normal parameters. This vertical lift must overcome gravity to properly center the rotating element in its bearing-support structure. However, because gravity and atmospheric pressure vary with altitude and barometric pressure, actual lift may not compensate for the downward forces of gravity in certain environments. When the deviation of actual lift from designed lift is significant, a rotor might not rotate on its true centerline. This offset rotation creates an imbalance and a measurable level of vibration.

Flow Instability and Operating Conditions

Rotating machines subject to imbalance caused by turbulent or unbalanced media flow include pumps, fans, and compressors. A good machine design for these units incorporates the dynamic forces of the gas or liquid in stabilizing the rotating element. The combination of these forces and the stiffness of the rotor-support system (i.e., bearing and bearing pedestals) determine the vibration level. Rotor-support stiffness is important because unbalanced forces resulting from flow instability can deflect rotating elements from their true centerline, and the stiffness resists the deflection.

Deviations from a machine's designed operating envelope can affect flow stability, which directly affects the vibration profile. For example, the vibration level of a centrifugal compressor is typically low when operating at 100% load with laminar airflow through the compressor. However, a radical change in vibration level can result from decreased load. Vibration resulting from operation at 50% load may increase by as much as 400% with no change in the mechanical condition of the compressor. In addition, a radical change in vibration level can result from turbulent flow caused by restrictions in either the inlet or discharge piping.

Turbulent or unbalanced media flow (i.e., aerodynamic or hydraulic instability) does not have the same quadratic impacts on the vibration profile as that of load change, but it increases the overall vibration energy. This generates a unique profile that can be used to quantify the level of instability present in the machine. The profile generated by unbalanced flow is visible at the vane or blade-pass frequency of the rotating element. In addition, the profile shows a marked increase in the random noise generated by the flow of gas or liquid through the machine.

Mechanical Motion and Forces

A clear understanding of the mechanical movement of machines and their components is an essential part of vibration analysis. This understanding, coupled with the forces applied by the process, are the foundation for diagnostic accuracy.

Almost every unique frequency contained in the vibration signature of a machine-train can be directly attributed to a corresponding mechanical motion within the machine. For example, the constant end play or axial movement of the rotating element in a motor-generator set generates an elevated amplitude at the fundamental (1x), second harmonic (2x), and third harmonic (3x) of the shaft's true running speed. In addition, this movement increases the axial amplitude of the fundamental (1x) frequency.

Forces resulting from air or liquid movement through a machine also generate unique frequency components within the machine's signature. In relatively stable or laminar-flow applications, the movement of product through the machine slightly increases the amplitude at the vane or blade-pass frequency. In more severe, turbulent-flow applications, the flow of product generates a broadband, white noise profile that can be directly attributed to the movement of product through the machine.

Other forces, such as the side-load created by V-belt drives, also generate unique frequencies or modify existing component frequencies. For example, excessive belt tension increases the side-load on the machine-train's shafts. This increase in side-load changes the load zone in the machine's bearings. The result of this change is a marked increase in the amplitude at the outer-race rotational frequency of the bearings.

Applied force or induced loads can also displace the shafts in a machine-train. As a result the machine's shaft will rotate off-center which dramatically increases the amplitude at the fundamental (1x) frequency of the machine.

RECIPROCATING AND/OR LINEAR-MOTION MACHINERY

This section describes machinery that exhibits reciprocating and/or linear motion(s) and discusses typical vibration behavior for these types of machines.

Machine Descriptions

Reciprocating linear-motion machines incorporate components that move linearly in a reciprocating fashion to perform work. Such reciprocating machines are bidirectional in that the linear movement reverses, returning to the initial position with each completed cycle of operation. Nonreciprocating linear-motion machines incorporate components that also generate work in a straight line, but do not reverse direction within one complete cycle of operation.

Few machines involve linear reciprocating motion exclusively. Most incorporate a combination of rotating and reciprocating linear motions to produce work. One example of such a machine is a reciprocating compressor. This unit contains a rotating crankshaft that transmits power to one or more reciprocating pistons, which move linearly in performing the work required to compress the media.

Sources of Vibration

Like rotating machinery, the vibration profile generated by reciprocating and/or linear-motion machines is the result of mechanical movement and forces generated by the components that are part of the machine. Vibration profiles generated by most reciprocating and/or linear-motion machines reflect a combination of rotating and/or linear-motion forces.

However, the intervals or frequencies generated by these machines are not always associated with one complete revolution of a shaft. In a two-cycle reciprocating engine, the pistons complete one cycle each time the crankshaft completes one 360-degree revolution. In a four-cycle engine, the crank must complete two complete revolutions, or 720 degrees, in order to complete a cycle of all pistons.

Because of the unique motion of reciprocating and linear-motion machines, the level of unbalanced forces generated by these machines is substantially higher than those generated by rotating machines. As an example, a reciprocating compressor drives each of its pistons from bottom-center to top-center and returns to bottom-center in each complete operation of the cylinder. The mechanical forces generated by the reversal of direction at both top-center and bottom-center result in a sharp increase in the vibration energy of the machine. An instantaneous spike in the vibration profile repeats each time the piston reverses direction.

Linear-motion machines generate vibration profiles similar to those of reciprocating machines. The major difference is the impact that occurs at the change of direction with reciprocating machines. Typically, linear-motion-only machines do not reverse direction during each cycle of operation and, as a result, do not generate the spike of energy associated with direction reversal.

Chapter 5

VIBRATION THEORY

Mathematical techniques allow us to quantify total displacement caused by all vibrations, to convert the displacement measurements to velocity or acceleration, to separate these data into their components through the use of FFT analysis, and to determine the amplitudes and phases of these functions. Such quantification is necessary if we are to isolate and correct abnormal vibrations in machinery.

PERIODIC MOTION

Vibration is a periodic motion, or one that repeats itself after a certain interval of time called the period, T. Figure 3.1 illustrated the periodic motion time-domain curve of a steam turbine bearing pedestal. Displacement is plotted on the vertical, or Y-axis, and time on the horizontal, or X-axis. The curve shown in Figure 3.4 is the sum of all vibration components generated by the rotating element and bearing-support structure of the turbine.

Harmonic Motion

The simplest kind of periodic motion or vibration, shown in Figure 3.2, is referred to as harmonic. Harmonic motions repeat each time the rotating element or machine component completes one complete cycle.

The relation between displacement and time for harmonic motion may be expressed by:

$$X = X_0 \sin(\omega t)$$

The maximum value of the displacement is X_0, which is also called the amplitude. The period, T, is usually measured in seconds; its reciprocal is the frequency of the vibration, f, measured in cycles-per-second (cps) or Hertz (Hz).

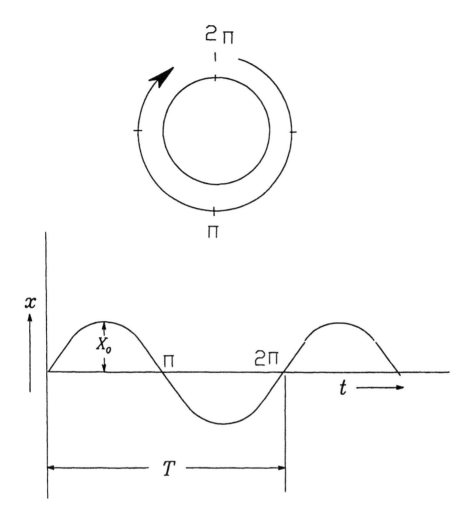

Figure 5.1 Illustration of vibration cycles.

$$f = \frac{1}{T}$$

Another measure of frequency is the circular frequency, ω, measured in radians per second. From Figure 5.1, it is clear that a full cycle of vibration (ωt) occurs after 360 degrees or 2π radians (i.e., one full revolution). At this point, the function begins a new cycle.

$$\omega = 2\pi f$$

For rotating machinery, the frequency is often expressed in vibrations per minute (vpm) or

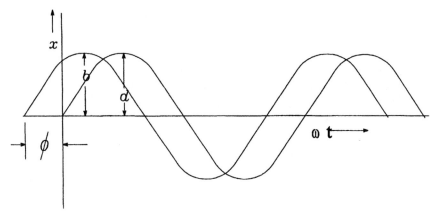

Figure 5.2 Two harmonic motions with a phase angle between them.

$$VPM = \frac{\omega}{\pi}$$

By definition, velocity is the first derivative of displacement with respect to time. For a harmonic motion, the displacement equation is:

$$X = X_0 \sin(\omega t)$$

The first derivative of this equation gives us the equation for velocity:

$$v = \frac{dX}{dt} = \dot{X} = \omega X_0 \cos(\omega t)$$

This relationship tells us that the velocity is also harmonic if the displacement is harmonic and has a maximum value or amplitude of $-\omega X_0$.

By definition, acceleration is the second derivative of displacement (i.e., the first derivative of velocity) with respect to time:

$$a = \frac{d^2 X}{dt^2} = \ddot{X} = -\omega^2 X_0 \sin(\omega t)$$

This function is also harmonic with amplitude of $\omega^2 X_0$.

Consider two frequencies given by the expression $X_1 = a \sin(\omega t)$ and $X_2 = b \sin(\omega t + \phi)$, which are shown in Figure 5.2 plotted against ωt as the X-axis. The quantity, ϕ, in the equation for X_2 is known as the phase angle or phase difference between the two vibrations. Because of ϕ, the two vibrations do not attain their maximum displacements at the same time. One is $\frac{\phi}{\omega}$ seconds behind the other. Note that

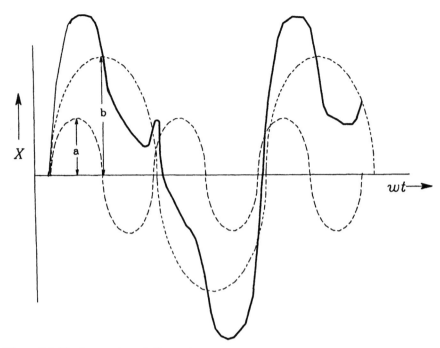

Figure 5.3 Nonharmonic periodic motion.

these two motions have the same frequency, ω. A phase angle has meaning only for two motions of the same frequency.

Nonharmonic Motion

In most machinery, there are numerous sources of vibrations, therefore, most time-domain vibration profiles are nonharmonic (represented by the solid line in Figure 5.3). While all harmonic motions are periodic, not every periodic motion is harmonic. Figure 5.3 is the superposition of two sine waves having different frequencies, and the dashed lines represent harmonic motions. These curves are represented by the following equations:

$$X_1 = a\sin(\omega_1 t)$$
$$X_2 = b\sin(\omega_2 t)$$

The total vibration represented by the solid line is the sum of the dashed lines. The following equation represents the total vibration:

$$X = X_1 + X_2 = a\sin(\omega_1 t) + b\sin(\omega_2 t)$$

Any periodic function can be represented as a series of sine functions having frequencies of ω, 2ω, 3ω, etc.:

$$f(t) = A_0 + A_1 \sin(\omega t + \phi_1) + A_2 \sin(2\omega t + \phi_2) + A_3 \sin(3\omega t + \phi_3) + \dots$$

This equation is known as a Fourier series, which is a function of time or $f(t)$. The amplitudes (A_1, A_2, etc.) of the various discrete vibrations and their phase angles ($\phi_1, \phi_2, \phi_3 \dots$) can be determined mathematically when the value of function $f(t)$ is known. Note that these data are obtained through the use of a transducer and a portable vibration analyzer.

The terms, 2ω, 3ω, etc., are referred to as the harmonics of the primary frequency, ω. In most vibration signatures, the primary frequency component is one of the running speeds of the machine-train (1x or 1ω). In addition, a signature may be expected to have one or more harmonics, for example, at two times (2x), three times (3x), and other multiples of the primary running speed.

MEASURABLE PARAMETERS

As shown previously, vibrations can be displayed graphically as plots, which are referred to as vibration profiles or signatures. These plots are based on measurable parameters (i.e., frequency and amplitude). Note that the terms *profile* and *signature* are sometimes used interchangeably by industry. In this module, however, *profile* is used to refer either to time-domain (also may be called time trace or waveform) or frequency-domain plots. The term *signature* refers to a frequency-domain plot.

Frequency

Frequency is defined as the number of repetitions of a specific forcing function or vibration component over a specific unit of time. Take for example a four-spoke wheel with an accelerometer attached. Every time the shaft completes one rotation, each of the four spokes passes the accelerometer once, which is referred to as four cycles per revolution. Therefore, if the shaft rotates at 100 rpm, the frequency of the spokes passing the accelerometer is 400 cycles per minute (cpm). In addition to cpm, frequency is commonly expressed in cycles per second (cps) or Hertz (Hz).

Note that for simplicity, a machine element's vibration frequency is commonly expressed as a multiple of the shaft's rotation speed. In the preceding example, the frequency would be indicated as 4X, or four times the running speed. In addition, because some malfunctions tend to occur at specific frequencies, it helps to segregate certain classes of malfunctions from others.

Note, however, that the frequency/malfunction relationship is not mutually exclusive and a specific mechanical problem cannot definitely be attributed to a unique frequency. While frequency is a very important piece of information with regard to isolating machinery malfunctions, it is only one part of the total picture. It is necessary to evaluate all data before arriving at a conclusion.

Amplitude

Amplitude refers to the maximum value of a motion or vibration. This value can be represented in terms of displacement (mils), velocity (inches per second), or acceleration (inches per second squared), each of which is discussed in more detail in the following section on Maximum Vibration Measurement.

Amplitude can be measured as the sum of all the forces causing vibrations within a piece of machinery (broadband), as discrete measurements for the individual forces (component), or for individual user-selected forces (narrowband). Broadband, component, and narrowband are discussed in a later section titled Measurement Classifications. Also discussed in this section are the common curve elements: peak-to-peak, zero-to-peak, and root-mean-square.

Maximum Vibration Measurement

The maximum value of a vibration, or amplitude, is expressed as displacement, velocity, or acceleration. Most of the microprocessor-based, frequency-domain vibration systems will convert the acquired data to the desired form. Because industrial vibration-severity standards are typically expressed in one of these terms, it is necessary to have a clear understanding of their relationship.

Displacement

Displacement is the actual change in distance or position of an object relative to a reference point and is usually expressed in units of mils, 0.001 inch. For example, displacement is the actual radial or axial movement of the shaft in relation to the normal centerline usually using the machine housing as the stationary reference. Vibration data, such as shaft displacement measurements acquired using a proximity probe or displacement transducer should always be expressed in terms of mils, peak-to-peak.

Velocity

Velocity is defined as the time rate of change of displacement (i.e., the first derivative, $\frac{dX}{dt}$ or $\dot{X}$) and is usually expressed as inches per second (in./sec). In simple terms, velocity is a description of how fast a vibration component is moving rather than how far, which is described by displacement.

Used in conjunction with zero-to-peak (PK) terms, velocity is the best representation of the true energy generated by a machine when relative or bearing cap data are used. (*Note:* Most vibration monitoring programs rely on data acquired from machine housing or bearing caps.) In most cases, peak velocity values are used with vibration data between 0 and 1000 Hz. These data are acquired with microprocessor-based, frequency-domain systems.

Acceleration

Acceleration is defined as the time rate of change of velocity (i.e., second derivative of displacement, $\dfrac{d^2 X}{dt^2}$ or $\ddot{X}$) and is expressed in units of inches per second squared (in./ sec²). Vibration frequencies above 1000 Hz should always be expressed as acceleration.

Acceleration is commonly expressed in terms of the gravitational constant, *g*, which is 32.17 ft/sec². In vibration analysis applications, acceleration is typically expressed in terms of *g*-RMS or *g*-PK. These are the best measures of the force generated by a machine, a group of components, or one of its components.

Measurement Classifications

At least three classifications of amplitude measurements are used in vibration analysis: broadband, narrowband, and component.

Broadband or Overall

The total energy of all vibration components generated by a machine is reflected by broadband, or overall, amplitude measurements. The normal convention for expressing the frequency range of broadband energy is a filtered range between 10 and 10,000 Hz, or 600 and 600,000 cpm. Because most vibration-severity charts are based on this filtered broadband, caution should be exercised to ensure that collected data are consistent with the charts.

Narrowband

Narrowband amplitude measurements refer to those that result from monitoring the energy generated by a user-selected group of vibration frequencies. Generally, this amplitude represents the energy generated by a filtered band of vibration components, failure mode, or forcing functions. For example, the total energy generated by flow instability can be captured using a filtered narrowband around the vane or blade-passing frequency.

Component

The energy generated by a unique machine component, motion, or other forcing function can yield its own amplitude measurement. For example, the energy generated by the rotational speed of a shaft, gear set meshing, or similar machine components generate discrete vibration components and their amplitude can be measured.

Common Elements of Curves

All vibration amplitude curves, which can represent displacement, velocity, or acceleration, have common elements that can be used to describe the function. These common elements are peak-to-peak, zero-to-peak, and root-mean-square, each of which is illustrated in Figure 5.4.

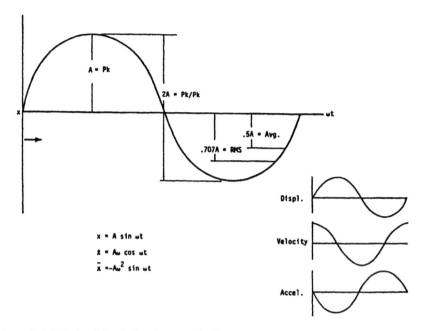

Figure 5.4 Relationship of vibration amplitude.

Peak-to-Peak

As illustrated in Figure 5.4, the peak-to-peak amplitude (2A, where A is the zero-to-peak) reflects the total amplitude generated by a machine, a group of components, or one of its components. This depends on whether the data gathered are broadband, narrowband, or component. The unit of measurement is useful when the analyst needs to know the total displacement or maximum energy produced by the machine's vibration profile.

Technically, peak-to-peak values should be used in conjunction with actual shaft-displacement data, which are measured with a proximity or displacement transducer. Peak-to-peak terms should not be used for vibration data acquired using either relative vibration data from bearing caps or when using a velocity or acceleration transducer. The only exception is when vibration levels must be compared to vibration-severity charts based on peak-to-peak values.

Zero-to-Peak

Zero-to-peak (A), or simply peak, values are equal to one-half of the peak-to-peak value. In general, relative vibration data acquired using a velocity transducer are expressed in terms of peak.

Root-Mean-Square

Root-mean-square (RMS) is the statistical average value of the amplitude generated by a machine, one of its components, or a group of components. Referring to Figure 5.4, RMS is equal to 0.707 of the zero-to-peak value, A. Normally, RMS data are used in conjunction with relative vibration data acquired using an accelerometer or expressed in terms of acceleration.

Chapter 6

MACHINE DYNAMICS

The primary reasons for vibration profile variations are the dynamics of the machine, which are affected by mass, stiffness, damping, and degrees of freedom. However, care must be taken because the vibration profile and energy levels generated by a machine also may vary depending on the location and orientation of the measurement.

MASS, STIFFNESS, AND DAMPING

The three primary factors that determine the normal vibration energy levels and the resulting vibration profiles are mass, stiffness, and damping. Every machine-train is designed with a dynamic support system that is based on the following: the mass of the dynamic component(s), a specific support system stiffness, and a specific amount of damping.

Mass

Mass is the property that describes how much material is present. Dynamically, it is the property that describes how an unrestricted body resists the application of an external force. Simply stated, the greater the mass the greater the force required to accelerate it. Mass is obtained by dividing the weight of a body (e.g., rotor assembly) by the local acceleration of gravity, g.

The English system of units is complicated compared to the metric system. In the English system, the units of mass are pounds-mass (lbm) and the units of weight are pounds-force (lbf). By definition, a weight (i.e., force) of 1 lbf equals the force produced by 1 lbm under the acceleration of gravity. Therefore, the constant, g_c, which has the same numerical value as g (32.17) and units of lbm-ft/lbf-sec^2, is used in the definition of weight:

$$Weight = \frac{Mass*g}{g_c}$$

Therefore,

$$Mass = \frac{Weight*g_c}{g}$$

Therefore,

$$Mass = \frac{Weight*g_c}{g} = \frac{lbf}{\frac{ft}{sec^2}} \times \frac{lbm*ft}{lbf*sec^2} = lbm$$

Stiffness

Stiffness is a spring-like property that describes the level of resisting force that results when a body undergoes a change in length. Units of stiffness are often given as pounds per inch (lbf/in.). Machine-trains have more than one stiffness property that must be considered in vibration analysis: shaft stiffness, vertical stiffness, and horizontal stiffness.

Shaft Stiffness

Most machine-trains used in industry have flexible shafts and relatively long spans between bearing-support points. As a result, these shafts tend to flex in normal operation. Three factors determine the amount of flex and mode shape that these shafts have in normal operation: shaft diameter, shaft material properties, and span length. A small-diameter shaft with a long span will obviously flex more than one with a larger diameter or shorter span.

Vertical Stiffness

The rotor-bearing support structure of a machine typically has more stiffness in the vertical plane than in the horizontal plane. Generally, the structural rigidity of a bearing-support structure is much greater in the vertical plane. The full weight of and the dynamic forces generated by the rotating element are fully supported by a pedestal cross-section that provides maximum stiffness.

In typical rotating machinery, the vibration profile generated by a normal machine contains lower amplitudes in the vertical plane. In most cases, this lower profile can be directly attributed to the difference in stiffness of the vertical plane when compared to the horizontal plane.

Horizontal Stiffness

Most bearing pedestals have more freedom in the horizontal direction than in the vertical. In most applications, the vertical height of the pedestal is much greater than the horizontal cross-section. As a result, the entire pedestal can flex in the horizontal plane as the machine rotates.

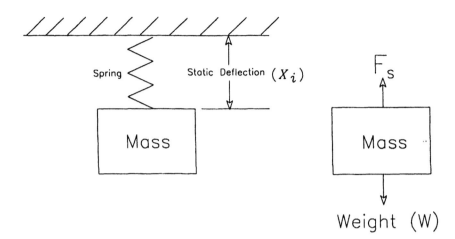

Figure 6.1 Undamped spring-mass system.

This lower stiffness generally results in higher vibration levels in the horizontal plane. This is especially true when the machine is subjected to abnormal modes of operation or when the machine is unbalanced or misaligned.

Damping

Damping is a means of reducing velocity through resistance to motion, in particular by forcing an object through a liquid or gas, or along another body. Units of damping are often given as pounds per inch per second (lbf/in./sec, which is also expressed as lbf-sec/in.).

The boundary conditions established by the machine design determine the freedom of movement permitted within the machine-train. A basic understanding of this concept is essential for vibration analysis. Free vibration refers to the vibration of a damped (as well as undamped) system of masses with motion entirely influenced by their potential energy. Forced vibration occurs when motion is sustained or driven by an applied periodic force in either damped or undamped systems. The following sections discuss free and forced vibration for both damped and undamped systems.

Free Vibration—Undamped

To understand the interactions of mass and stiffness, consider the case of undamped free vibration of a single mass that only moves vertically, as illustrated in Figure 6.1. In this figure, the mass M is supported by a spring that has a stiffness K (also referred to as the spring constant), which is defined as the number of pounds of tension necessary to extend the spring 1 in.

The force created by the static deflection, X_i, of the spring supports the weight, W, of the mass. Also included in Figure 6.1 is the free-body diagram that illustrates the two forces acting on the mass. These forces are the weight (also referred to as the inertia force) and an equal, yet opposite force that results from the spring (referred to as the spring force, F_s).

The relationship between the weight of mass M and the static deflection of the spring can be calculated using the following equation:

$$W = KX_i$$

If the spring is displaced downward some distance, X_0, from X_i and released, it will oscillate up and down. The force from the spring, F_s, can be written as follows, where a is the acceleration of the mass:

$$F_s = -KX = \frac{Ma}{g_c}$$

It is common practice to replace acceleration a with $\dfrac{d^2X}{dt^2}$, the second derivative of the displacement, X, of the mass with respect to time, t. Making this substitution, the equation that defines the motion of the mass can be expressed as:

$$\frac{M}{g_c}\frac{d^2X}{dt^2} = -KX \quad \text{or} \quad \frac{M}{g_c}\frac{d^2X}{dt^2} + KX = 0$$

Motion of the mass is known to be periodic in time. Therefore, the displacement can be described by the expression:

$$X = X_0\cos(\omega t)$$

where

X = Displacement at time t
X_0 = Initial displacement of the mass
ω = Frequency of the oscillation (natural or resonant frequency)
t = Time.

If this equation is differentiated and the result inserted into the equation that defines motion, the natural frequency of the mass can be calculated. The first derivative of the equation for motion given previously yields the equation for velocity. The second derivative of the equation yields acceleration.

$$Velocity = \frac{dX}{dt} = \dot{X} = -\omega X_0\sin(\omega t)$$

$$Acceleration = \frac{d^2X}{dt^2} = \ddot{X} = -\omega^2 X_0 \cos(\omega t)$$

Inserting the above expression for acceleration, or $\frac{d^2X}{dt^2}$, into the equation for F_s yields the following:

$$\frac{M}{g_c}\frac{d^2X}{dt^2} + KX = 0$$

$$-\frac{M}{g_c}\omega^2 X_0 \cos(\omega t) + KX = 0$$

$$-\frac{M}{g_c}\omega^2 X + KX = -\frac{M}{g_c}\omega^2 + K = 0$$

Solving this expression for ω yields the equation:

$$\omega = \sqrt{\frac{K g_c}{M}}$$

where

ω = Natural frequency of mass
K = Spring constant
M = Mass.

Note that, theoretically, undamped free vibration persists forever. However, this never occurs in nature and all free vibrations die down after time due to damping, which is discussed in the next section.

Free Vibration—Damped

A slight increase in system complexity results when a damping element is added to the spring-mass system shown in Figure 6.2. This type of damping is referred to as viscous damping. Dynamically, this system is the same as the undamped system illustrated in Figure 6.1, except for the damper, which usually is an oil or air dashpot mechanism. A damper is used to continuously decrease the velocity and the resulting energy of a mass undergoing oscillatory motion.

The system is still comprised of the inertia force due to the mass and the spring force, but a new force is introduced. This force is referred to as the damping force and is proportional to the damping constant, or the coefficient of viscous damping, c. The damping force is also proportional to the velocity of the body and, as it is applied, it opposes the motion at each instant.

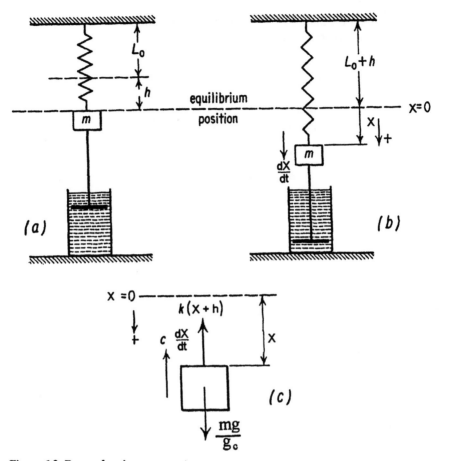

Figure 6.2 Damped spring-mass system.

In Figure 6.2, the unelongated length of the spring is L_0 and the elongation due to the weight of the mass is expressed by h. Therefore, the weight of the mass is Kh. Figure 6.2(a) shows the mass in its position of stable equilibrium. Figure 6.2(b) shows the mass displaced downward a distance X from the equilibrium position. Note that X is considered positive in the downward direction.

Figure 6.2(c) is a free-body diagram of the mass, which has three forces acting on it. The weight (Mg/g_c), which is directed downward, is always positive. The damping force $\left(c \dfrac{dX}{dt} \right)$, which is the damping constant times velocity, acts opposite to the direction of the velocity. The spring force, $K(X + h)$, acts in the direction opposite to the displacement. Using Newton's equation of motion, where $\sum F = Ma$, the sum of the forces acting on the mass can be represented by the following equation, remembering that X is positive in the downward direction:

$$\frac{M}{g_c} \frac{d^2X}{dt^2} = \frac{Mg}{g_c} - c\frac{dX}{dt} - K(X+h)$$

$$\frac{M}{g_c} \frac{d^2X}{dt^2} = Kh - c\frac{dX}{dt} - KX - Kh$$

$$\frac{M}{g_c} \frac{d^2X}{dt^2} = -c\frac{dX}{dt} - KX$$

Dividing by $\dfrac{M}{g_c}$:

$$\frac{d^2X}{dt^2} = -\frac{cg_c}{M}\frac{dX}{dt} - \frac{Kg_c X}{M}$$

To look up the solution to the preceding equation in a differential equations table (such as in the *CRC Handbook of Chemistry and Physics*) it is necessary to change the form of this equation. This can be accomplished by defining the relationships, $cg_c /M = 2\mu$ and $Kg_c /M = \omega^2$, which converts the equation to the following form:

$$\frac{d^2X}{dt^2} = -2\mu\frac{dX}{dt} - \omega^2 X$$

Note that for undamped free vibration, the damping constant, c, is zero and, therefore, μ is also zero.

$$\frac{d^2X}{dt^2} = -\omega^2 X$$

$$\frac{d^2X}{dt^2} + \omega^2 X = 0$$

The solution of this equation describes simple harmonic motion, which is given below:

$$X = A\cos(\omega t) + B\sin(\omega t)$$

Substituting at $t = 0$, then $X = X_0$ and $\dfrac{dX}{dt} = 0$, then

$$X = X_0 \cos(\omega t)$$

This shows that free vibration is periodic and is the solution for X. For damped free vibration, however, the damping constant, c, is not zero.

$$\frac{d^2X}{dt^2} = -2\mu\frac{dX}{dt} - \omega^2 X$$

or

$$\frac{d^2X}{dt^2} + 2\mu\frac{dX}{dt} + \omega^2 X = 0$$

or

$$D^2 + 2\mu D + \omega^2 = 0$$

which has a solution of:

$$X = Ae^{d^1 t} + Be^{d^2 t}$$

where

$$d_1 = -\mu + \sqrt{\mu^2 - \omega^2}$$

$$d_2 = -\mu - \sqrt{\mu^2 - \omega^2}$$

There are different conditions of damping: critical, overdamping, and underdamping. Critical damping occurs when $\mu = \omega$. Overdamping occurs when $\mu > \omega$. Underdamping occurs when $\mu < \omega$.

The only condition that results in oscillatory motion and, therefore, represents a mechanical vibration is underdamping. The other two conditions result in aperiodic motions. When damping is less than critical ($\mu < \omega$), then the following equation applies:

$$X = \frac{X_0}{\alpha_1}e^{-\mu t}(\alpha_1\cos\alpha_1 t + \mu\sin\alpha_1 t)$$

where

$$\alpha_1 = \sqrt{\omega^2 - \mu^2}$$

Forced Vibration—Undamped

The simple systems described in the preceding two sections on free vibration are alike in that they are not forced to vibrate by any exciting force or motion. Their major contribution to the discussion of vibration fundamentals is that they illustrate how a system's natural or resonant frequency depends on the mass, stiffness, and damping characteristics.

The mass–stiffness–damping system also can be disturbed by a periodic variation of external forces applied to the mass at any frequency. The system shown in Figure 6.1 is increased in complexity by the addition of an external force, F_0, acting downward on the mass.

In undamped forced vibration, the only difference in the equation for undamped free vibration is that instead of the equation being equal to zero, it is equal to $F_0 \sin(\omega t)$:

$$\frac{M}{g_c} \frac{d^2 X}{dt^2} + KX = F_0 \sin(\omega t)$$

Since the spring is not initially displaced and is "driven" by the function $F_0 \sin(\omega t)$, a particular solution, $X = X_0 \sin(\omega t)$, is logical. Substituting this solution into the above equation and performing mathematical manipulations yields the following equation for X:

$$X = C_1 \sin(\omega_n t) + C_2 \cos(\omega_n t) + \frac{X_{st}}{1 - (\omega/\omega_n)^2} \sin(\omega t)$$

where

X = Spring displacement at time, t
X_{st} = Static spring deflection under constant load, F_0
ω = Forced frequency
ω_n = Natural frequency of the oscillation
t = Time
C_1, C_2 = Integration constants determined from specific boundary conditions.

In the above equation, the first two terms are the undamped free vibration, and the third term is the undamped forced vibration. The solution, containing the sum of two sine waves of different frequencies, is itself not a harmonic motion.

Forced Vibration—Damped

In a damped forced vibration system such as the one shown in Figure 6.3, the motion of the mass M has two parts: (1) the damped free vibration at the damped natural frequency and (2) the steady-state harmonic motions at the forcing frequency. The damped natural frequency component decays quickly, but the steady-state harmonic associated with the external force remains as long as the energy force is present.

With damped forced vibration, the only difference in its equation and the equation for damped free vibration is that it is equal to $F_0 \sin(\omega t)$ as shown below instead of being equal to zero.

$$\frac{M}{g_c} \frac{d^2 X}{dt^2} + c\frac{dX}{dt} + KX = F_0 \sin(\omega t)$$

With damped vibration, damping constant c is not equal to zero and the solution of the equation gets quite complex assuming the function, $X = X_0 \sin(\omega t - \phi)$. In this equation, ϕ is the phase angle, or the number of degrees that the external force, $F_0 \sin(\omega t)$, is ahead of the displacement, $X_0 \sin(\omega t - \phi)$. Using vector concepts, the following

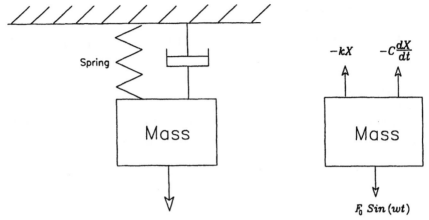

Figure 6.3 Damped forced vibration system.

equations apply, which can be solved because there are two equations and two unknowns:

Vertical vector component: $KX_0 - \dfrac{M}{g_c}\omega^2 X_0 - F_0\cos\phi = 0$

Horizontal vector component: $c\omega X_0 - F_0\sin\phi = 0$

Solving these two equations for the unknowns X_0 and ϕ:

$$X_0 = \frac{F_0}{\sqrt{(c\omega)^2 + \left(K - \dfrac{M}{g_c}\omega^2\right)^2}} = \frac{\dfrac{F_0}{K}}{\sqrt{\left(1 - \dfrac{\omega^2}{\omega_n^2}\right) + \left(2\dfrac{c}{c_c} \times \dfrac{\omega}{\omega_n}\right)^2}}$$

$$\tan\phi = \frac{c\omega}{K - \dfrac{M}{g_c}\omega^2} = \frac{2\dfrac{c}{c_c} \times \dfrac{\omega}{\omega_n}}{1 - (\omega^2/\omega_n^2)}$$

where

c = Damping constant

c_c = Critical damping = $2\dfrac{M}{g_c}\omega_n$

c/c_c = Damping ratio

F_0 = External force

F_0/K = Deflection of the spring under load, F_0 (also called static deflection, X_{st})

ω = Forced frequency

ω_n = Natural frequency of the oscillation

ω/ω_n = Frequency ratio.

For damped forced vibrations, three different frequencies have to be distinguished: the undamped natural frequency, $\omega_n = \sqrt{Kg_c/M}$; the damped natural frequency,

$q = \sqrt{\dfrac{Kg_c}{M} - \left(\dfrac{cg_c}{2M}\right)^2}$; and the frequency of maximum forced amplitude, sometimes referred to as the resonant frequency.

DEGREES OF FREEDOM

In a mechanical system, the degrees of freedom indicate how many numbers are required to express its geometrical position at any instant. In machine-trains, the relationship of mass, stiffness, and damping is not the same in all directions. As a result, the rotating or dynamic elements within the machine move more in one direction than in another. A clear understanding of the degrees of freedom is important in that it has a direct impact on the vibration amplitudes generated by a machine or process system.

One Degree of Freedom

If the geometrical position of a mechanical system can be defined or expressed as a single value, the machine is said to have one degree of freedom. For example, the position of a piston moving in a cylinder can be specified at any point in time by measuring the distance from the cylinder end.

A single degree of freedom is not limited to simple mechanical systems such as the cylinder. For example, a 12-cylinder gasoline engine with a rigid crankshaft and a rigidly mounted cylinder block has only one degree of freedom. The position of all of its moving parts (i.e., pistons, rods, valves, cam shafts, etc.) can be expressed by a single value. In this instance, the value would be the angle of the crankshaft.

However, when mounted on flexible springs, this engine has multiple degrees of freedom. In addition to the movement of its internal parts in relationship to the crank, the entire engine can now move in any direction. As a result, the position of the engine and any of its internal parts require more than one value to plot its actual position in space.

The definitions and relationships of mass, stiffness, and damping in the preceding section assumed a single degree of freedom. In other words, movement was limited to a

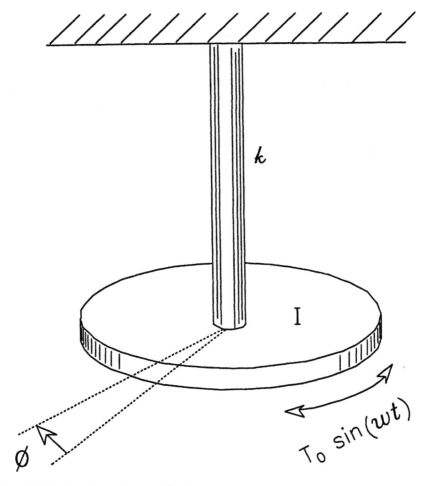

Figure 6.4 Torsional one-degree-of-freedom system.

single plane. Therefore, the formulas are applicable for all single-degree-of-freedom mechanical systems.

The calculation for torque is a primary example of a single degree of freedom in a mechanical system. Figure 6.4 represents a disk with a moment of inertia, I, that is attached to a shaft of torsional stiffness, k.

Torsional stiffness is defined as the externally applied torque, T, in inch-pounds needed to turn the disk one radian (57.3 degrees). Torque can be represented by the following equations:

$$\sum Torque = \text{Moment of inertia} \times \text{angular acceleration} = I\frac{d^2\phi}{dt^2} = I\ddot{\phi}$$

In this example, three torques are acting on the disk: the spring torque, damping torque (due to the viscosity of the air), and external torque. The spring torque is minus $(-)k\phi$, where ϕ is measured in radians. The damping torque is minus $(-)c\dot{\phi}$, where c is the damping constant. In this example, c is the damping torque on the disk caused by an angular speed of rotation of one radian per second. The external torque is $T_0 \sin(\omega t)$.

$$I\ddot{\phi} = \sum Torque = -c\dot{\phi} - k\phi + T_0\sin(\omega t)$$

or

$$I\ddot{\phi} + c\dot{\phi} + k\phi = T_0\sin(\omega t)$$

Two Degrees of Freedom

The theory for a one-degree-of-freedom system is useful for determining resonant or natural frequencies that occur in all machine-trains and process systems. However, few machines have only one degree of freedom. Practically, most machines will have two or more degrees of freedom. This section provides a brief overview of the theories associated with two degrees of freedom. An undamped two-degree-of-freedom system is illustrated in Figure 6.5.

The diagram of Figure 6.5 consists of two masses, M_1 and M_2, which are suspended from springs, K_1 and K_2. The two masses are tied together, or coupled, by spring K_3, so that they are forced to act together. In this example, the movement of the two masses is limited to the vertical plane and, therefore, horizontal movement can be ignored. As in the single-degree-of-freedom examples, the absolute position of each mass is defined by its vertical position above or below the neutral, or reference, point. Since there are two coupled masses, two locations (i.e., one for M_1 and one for M_2) are required to locate the absolute position of the system.

To calculate the free or natural modes of vibration, note that two distinct forces are acting on mass, M_1: the force of the main spring, K_1, and that of the coupling spring, K_3. The main force acts upward and is defined as $-K_1X_1$. The shortening of the coupling spring is equal to the difference in the vertical position of the two masses, $X_1 - X_2$. Therefore, the compressive force of the coupling spring is $K_3(X_1 - X_2)$. The compressed coupling spring pushes the top mass, M_1, upward so that the force is negative.

Because these are the only tangible forces acting on M_1, the equation of motion for the top mass can be written as:

$$\frac{M_1}{g_c}\ddot{X}_1 = -K_1X_1 - K_3(X_1 - X_2)$$

or

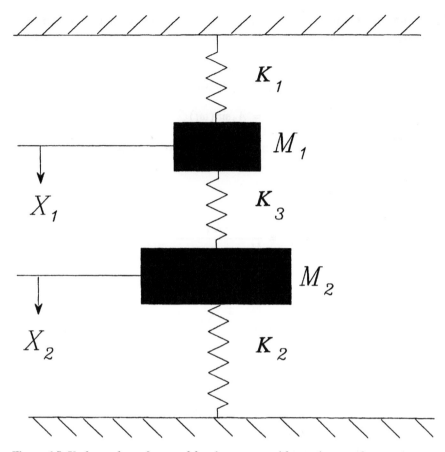

Figure 6.5 Undamped two-degree-of-freedom system with a spring couple.

$$\frac{M_1}{g_c}\ddot{X}_1 + (K_1 + K_3)X_1 - K_3X_2 = 0$$

The equation of motion for the second mass, M_2, is derived in the same manner. To make it easier to understand, turn the figure upside down and reverse the direction of X_1 and X_2. The equation then becomes:

$$\frac{M_2}{g_c}\ddot{X}_2 = -K_2X_2 - K_3(X_1 - X_2)$$

or

$$\frac{M_2}{g_c}\ddot{X}_2 + (K_2 + K_3)X_2 - K_3X_1 = 0$$

If we assume that the masses M_1 and M_2 undergo harmonic motions with the same frequency, ω, and with different amplitudes, A_1 and A_2, their behavior can be represented as follows:

$$X_1 = A_1 \sin(\omega t)$$

$$X_2 = A_2 \sin(\omega t)$$

By substituting these into the differential equations, two equations for the amplitude ratio, $\dfrac{A_1}{A_2}$, can be found:

$$\frac{A_1}{A_2} = \frac{-K_3}{\dfrac{M_1}{g_c}\omega^2 - K_1 - K_3}$$

$$\frac{A_1}{A_2} = \frac{\dfrac{M_2}{g_c}\omega^2 - K_2 - K_3}{-K_3}$$

For a solution of the form we assumed to exist, these two equations must be equal:

$$\frac{-K_3}{\dfrac{M_1}{g_c}\omega^2 - K_1 - K_3} = \frac{\dfrac{M_2}{g_c}\omega^2 - K_2 - K_3}{-K_3}$$

or

$$\omega^4 - \omega^2\left\{\frac{K_1 + K_3}{M_1/g_c} + \frac{K_2 + K_3}{M_2/g_c}\right\} + \frac{K_1 K_2 + K_2 K_3 + K_1 K_3}{\dfrac{M_1 M_2}{g_c^2}} = 0$$

This equation, known as the frequency equation, has two solutions for ω^2. When substituted in either of the preceding equations, each one of these gives a definite value for $\dfrac{A_1}{A_2}$. This means that there are two solutions for this example, which are of the form $A_1 \sin(\omega t)$ and $A_2 \sin(\omega t)$. As with many such problems, the final answer is the superposition of the two solutions with the final amplitudes and frequencies determined by the boundary conditions.

Many Degrees of Freedom

When the number of degrees of freedom becomes greater than two, no critical new parameters enter into the problem. The dynamics of all machines can be understood by following the rules and guidelines established in the one- and two-degree-of-freedom equations. There are as many natural frequencies and modes of motion as there are degrees of freedom.

Chapter 7

VIBRATION DATA TYPES AND FORMATS

There are several options regarding the types of vibration data that can be gathered for machine-trains and systems and the formats in which it can be collected. However, selection of type and format depends on the specific application.

The two major data-type classifications are time domain and frequency domain. Each of these can be further divided into steady-state and dynamic data formats. In turn, each of these two formats can be further divided into single-channel and multiple-channel formats.

DATA TYPES

Vibration profiles can be acquired and displayed in one of two data types: (1) time domain or (2) frequency domain.

Time-Domain Data

Most of the early vibration analyses were carried out using analog equipment, which necessitated the use of time-domain data. The reason for this is that it was difficult to convert time-domain data to frequency-domain data. Frequency-domain capability was not available until microprocessor-based analyzers incorporated a straightforward method (i.e., fast Fourier transform) for transforming the time-domain spectrum into its frequency components.

Actual time-domain vibration signatures are commonly referred to as time traces or time plots (see Figure 7.1). Theoretical vibration data are generally referred to as waveforms (see Figure 7.2).

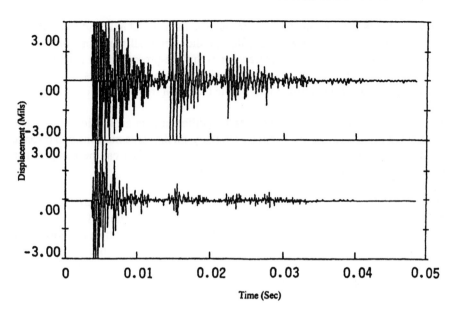

Figure 7.1 Typical time-domain signature.

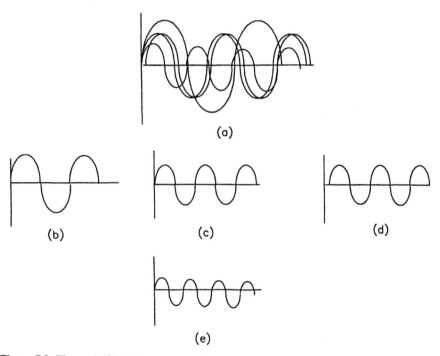

Figure 7.2 Theoretical time-domain waveforms.

Time-domain data are presented with amplitude as the vertical axis and elapsed time as the horizontal axis. Time-domain profiles are the sum of all vibration components (i.e., frequencies, impacts, and other transients) that are present in the machine-train and its installed system. Time traces include all frequency components, but the individual components are more difficult to isolate than with frequency-domain data.

The profile shown in Figure 7.2 illustrates two different data acquisition points, one measured vertically and one measured horizontally, on the same machine and taken at the same time. Because they were obtained concurrently, they can be compared to determine the operating dynamics of the machine.

In this example, the data set contains an impact that occurred at 0.005 sec. The impact is clearly visible in both the vertical (top) and horizontal (bottom) data sets. From these time traces, it is apparent that the vertical impact is stronger than the horizontal. In addition, the impact is repeated at 0.015 and 0.025 sec. Two conclusions can be derived from this example: (1) The impact source is a vertical force and (2) it impacts the machine-train at an interval of 0.010 sec, or a frequency of 1/0.010 sec = 100 Hz.

The waveform in Figure 7.2 illustrates theoretically the unique frequencies and transients that could be present in a machine's signature. The figure illustrates the complexity of such a waveform by overlaying numerous frequencies. The discrete waveforms that make up Figure 7.2 are displayed individually in Figures 7.2(b)–(e). Note that two of the frequencies, shown in Figures 7.2(c) and (d), are identical, but have a different phase angle (ϕ).

With time-domain data, the analyst must manually separate the individual frequencies and events that are contained in the complex waveform. This effort is complicated tremendously by the superposition of multiple frequencies. Note that, rather than overlaying each of the discrete frequencies as illustrated theoretically in Figure 7.2(a), actual time-domain data represent the sum of these frequencies as was illustrated in Figure 7.1.

To analyze this type of plot, the analyst must manually change the timescale to obtain discrete frequency curve data. The time interval between the recurrence of each frequency can then be measured. In this way, it is possible to isolate each of the frequencies that make up the time-domain vibration signature.

For routine monitoring of machine vibration, however, this approach is not cost effective. The time required to isolate manually each of the frequency components and transient events contained in the waveform is prohibitive. However, time-domain data have a definite use in a total plant predictive maintenance or reliability improvement program.

Machine-trains or process systems that have specific timing events (e.g., a pneumatic or hydraulic cylinder) must be analyzed using time-domain data format. In addition, time-domain data must be used for linear and reciprocating motion machinery.

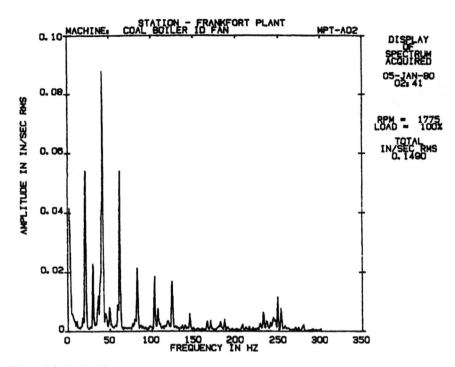

Figure 7.3 Typical frequency-domain signature.

Frequency-Domain Data

Most rotating machine-train failures result at or near a frequency component associated with the running speed. Therefore, the ability to display and analyze the vibration spectrum as components of frequency is extremely important.

The frequency-domain format eliminates the manual effort required to isolate the components that make up a time trace. Frequency-domain techniques convert time-domain data into discrete frequency components using a fast Fourier transform (FFT). Simply stated, FFT mathematically converts a time-based trace into a series of discrete frequency components (see Figure 7.3). In a frequency-domain plot, the X-axis is frequency and the Y-axis is the amplitude of displacement, velocity, or acceleration.

With frequency-domain analysis, the average spectrum for a machine-train signature can be obtained. Recurring peaks can be normalized to present an accurate representation of the machine-train condition. Figure 7.4 illustrates a simplified relationship between the time-domain and frequency-domain methods.

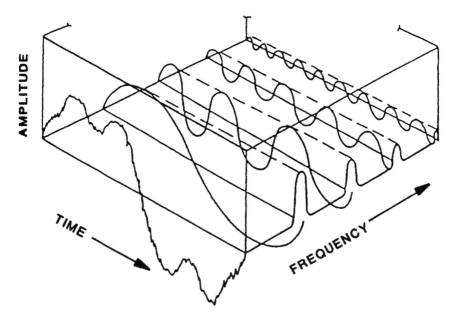

Figure 7.4 Relationship between time domain and frequency domain.

The real advantage of frequency-domain analysis is the ability to normalize each vibration component so that a complex machine-train spectrum can be divided into discrete components. This ability simplifies isolation and analysis of mechanical degradation within the machine-train. In addition, note that frequency-domain analysis can be used to determine the phase relationships for harmonic vibration components in a typical machine-train spectrum. The frequency domain normalizes any or all running speeds, whereas time-domain analysis is limited to true running speed.

Mathematical theory shows that any periodic function of time, $f(t)$, can be represented as a series of sine functions having frequencies ω, 2ω, 3ω, 4ω, etc. Function $f(t)$ is represented by the following equation, which is referred to as a Fourier series:

$$f(t) = A_0 + A_1 \sin(\omega t + \phi_1) + A_2 \sin(2\omega t + \phi_2) + A_3 \sin(3\omega t + \phi_3) + \dots ,$$

where

  A_x = Amplitude of each discrete sine wave
  ω = Frequency
  ϕ_x = Phase angle of each discrete sine wave.

Each of these sine functions represents a discrete component of the vibration signature discussed previously. The amplitudes of each discrete component and their phase angles can be determined by integral calculus when the function $f(t)$ is known. Because the subject of integral calculus is beyond the scope of this module, the math required to determine these integrals is not presented. A vibration analyzer and its associated software perform this determination using FFT.

DATA FORMATS

Both time-domain and frequency-domain vibration data can be acquired and analyzed in two primary formats: (1) steady state or (2) dynamic. Each of these formats has strengths and weaknesses that must be clearly understood for proper use. Each of these formats can be obtained as single- or multiple-channel data.

Steady-State Format

Most vibration programs that use microprocessor-based analyzers are limited to steady-state data. Steady-state vibration data assume the machine-train or process system operates in a constant, or steady-state, condition. In other words, the machine is free of dynamic variables such as load, flow, etc. This approach further assumes that all vibration frequencies are repeatable and maintain a constant relationship to the rotating speed of the machine's shaft.

Steady-state analysis techniques are based on acquiring vibration data when the machine or process system is operating at a fixed speed and specific operating parameters. For example, a variable-speed machine-train is evaluated at constant speed rather than over its speed range.

Steady-state analysis can be compared to a still photograph of the vibration profile generated by a machine or process system. Snapshots of the vibration profile are acquired by the vibration analyzer and stored for analysis. While the snapshots can be used to evaluate the relative operating condition of simple machine-trains, they do not provide a true picture of the dynamics of either the machine or its vibration profile.

Steady-state analysis totally ignores variations in the vibration level or vibration generated by transient events such as impacts and changes in speed or process parameters. Instruments used to obtain the profiles contain electronic circuitry, which are specifically designed to eliminate transient data.

In the normal acquisition process, the analyzer acquires multiple blocks of data. As part of the process, the microprocessor compares each block of data as it is acquired. If a block contains a transient that is not included in subsequent blocks, the block containing the event is discarded and replaced with a transient-free block. As a result, steady-state analysis does not detect random events that may have a direct, negative effect on equipment reliability.

Dynamic Format

While steady-state data provide a snapshot of the machine, dynamic or real-time data provide a motion picture. This approach provides a better picture of the dynamics of both the machine-train and its vibration profile. Data acquired using steady-state methods would suggest that vibration profiles and amplitudes are constant. However, this is not true. All dynamic forces, including running speed, vary constantly in all machine-trains. When real-time data acquisition methods are used, these variations are captured and displayed for analysis.

Single-Channel Format

Most microprocessor-based vibration monitoring programs rely on single-channel vibration data format. Single-channel data acquisition and analysis techniques are acceptable for routine monitoring of simple, rotating machinery. However, it is important that single-channel analysis be augmented with multiple-channel and dynamic analysis. Total reliance on single-channel techniques severely limits the accuracy of analysis and the effectiveness of a predictive maintenance or reliability improvement program.

With the single-channel method, data are acquired in series or one channel at a time. Normally, a series of data points is established for each machine-train, and data are acquired from each point in a measurement route. Whereas this approach is more than adequate for routine monitoring of relatively simple machines, it is based on the assumption that the machine's dynamics and the resultant vibration profile are constant throughout the entire data acquisition process. This approach hinders the ability to evaluate real-time relationships between measurement points on the machine-train and variations in process parameters such as speed, load, pressure, etc.

Multiple-Channel Format

Multiple-channel data provide the best picture of the relationship between measurement points on a machine-train. Data are acquired simultaneously from all measurement points on the machine-train. With this type of data, the analyst can establish the relationship between machine dynamics and vibration profile of the entire machine.

In most cases, a digital tape recorder is used to acquire data from the machine. Because all measurement points are recorded at the same time, the resultant data can be used to compare the triaxial vibration profile of all measurement points. This capability greatly enhances the analyst's ability to isolate abnormal machine dynamics and to determine the root cause of deviations.

Chapter 8

DATA ACQUISITION

It is important for predictive maintenance programs using vibration analysis to have accurate, repeatable data. In addition to the type and quality of the transducer, three key parameters affect data quality: the point of measurement, orientation, and transducer-mounting techniques.

In a predictive and reliability maintenance program, it is extremely important to keep good historical records of key parameters. How measurement point locations and orientation to the machine's shaft were selected should be kept as part of the database. It is important that every measurement taken throughout the life of the maintenance program be acquired at exactly the same point and orientation. In addition, the compressive load, or downward force, applied to the transducer should be exactly the same for each measurement.

VIBRATION DETECTORS: TRANSDUCERS AND CABLES

A variety of monitoring, trending, and analysis techniques are available that can and should be used as part of a total plant vibration monitoring program. Initially, such a program depends on the use of historical trends to detect incipient problems. As the program matures, however, other techniques such as frequency-domain signature analysis, time-domain analysis, and operating dynamics analysis are typically added.

An analysis is only as good as the data used, therefore, the equipment used to collect the data are critical and determine the success or failure of a predictive maintenance or reliability improvement program. The accuracy and proper use and mounting of equipment determines whether or not valid data are collected.

Specifically, three basic types of vibration transducers can be used for monitoring the mechanical condition of plant machinery: displacement probes, velocity transducers,

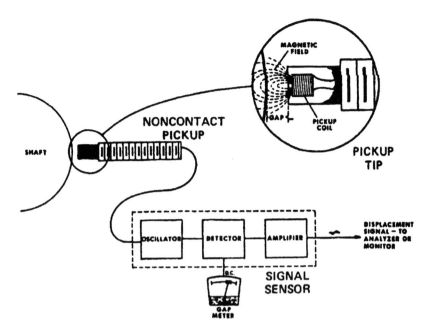

Figure 8.1 Displacement probe and signal conditioning system.

and accelerometers. Each has limitations and specific applications for which its use is appropriate.

Displacement Probes

Displacement, or eddy-current, probes are designed to measure the actual movement, or displacement, of a machine's shaft relative to the probe. Data are normally recorded as peak-to-peak in mils, or thousandths of an inch. This value represents the maximum deflection or displacement from the true centerline of a machine's shaft. Such a device must be rigidly mounted to a stationary structure to obtain accurate, repeatable data. Figure 8.1 shows an illustration of a displacement probe and signal conditioning system.

Permanently mounted displacement probes provide the most accurate data on machines having a rotor weight that is low relative to the casing and support structure. Turbines, large compressors, and other types of plant equipment should have displacement transducers permanently mounted at key measurement locations.

The useful frequency range for displacement probes is from 10 to 1000 Hz, or 600 to 60,000 rpm. Frequency components above or below this range are distorted and, therefore, unreliable for determining machine condition.

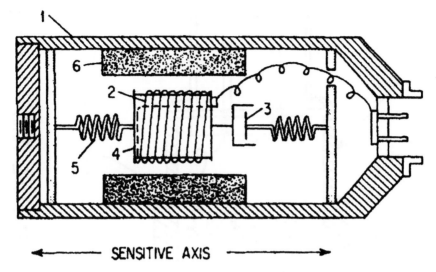

Figure 8.2 Schematic diagram of velocity pickup: (1) pickup case, (2) wire out, (3) damper, (4) mass, (5) spring, (6) magnet.

The major limitation with displacement or proximity probes is cost. The typical cost for installing a single probe, including a power supply, signal conditioning, etc., averages $1000. If each machine to be evaluated requires 10 measurements, the cost per machine is about $10,000. Using displacement transducers for all plant machinery dramatically increases the initial cost of the program. Therefore, key locations are generally instrumented first and other measurement points are added later.

Velocity Transducers

Velocity transducers are electromechanical sensors designed to monitor casing, or relative, vibration. Unlike displacement probes, velocity transducers measure the rate of displacement rather than the distance of movement. Velocity is normally expressed in terms of inches per second (in./sec) peak, which is perhaps the best method of expressing the energy caused by machine vibration. Figure 8.2 is a schematic diagram of a velocity measurement device.

Like displacement probes, velocity transducers have an effective frequency range of about 10 to 1000 Hz. They should not be used to monitor frequencies above or below this range.

The major limitation of velocity transducers is their sensitivity to mechanical and thermal damage. Normal use can cause a loss of calibration and, therefore, a strict recalibration program is required to prevent data errors. At a minimum, velocity transducers should be recalibrated every 6 months. Even with periodic recalibration, however, velocity transducers are prone to provide distorted data due to loss of calibration.

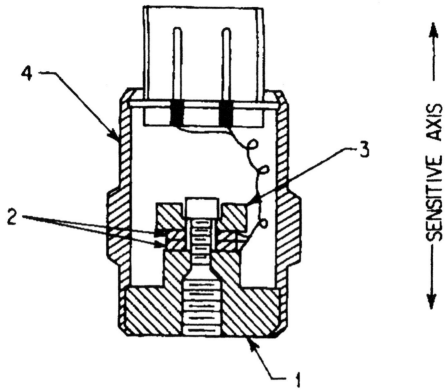

SENSITIVE AXIS

Figure 8.3 Schematic diagram of accelerometer: (1) base, (2) piezoelectric crystals, (3) mass, (4) case.

Accelerometers

Acceleration is perhaps the best method of determining the force resulting from machine vibration. Accelerometers use piezoelectric crystals or films to convert mechanical energy into electrical signals and Figure 8.3 is a schematic of such a device. Data acquired with this type of transducer are relative acceleration expressed in terms of the gravitational constant, g, in inches/second/second.

The effective range of general-purpose accelerometers is from about 1 to 10,000 Hz. Ultrasonic accelerometers are available for frequencies up to 1 MHz. In general, vibration data above 1000 Hz (or 60,000 cpm) should be taken and analyzed in acceleration or g's.

A benefit of the use of accelerometers is that they do not require a calibration program to ensure accuracy. However, they are susceptible to thermal damage. If sufficient heat radiates into the piezoelectric crystal, it can be damaged or destroyed. However,

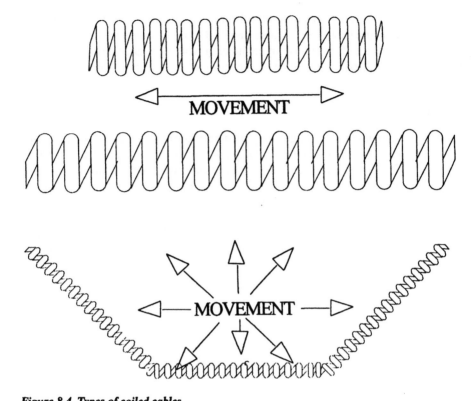

Figure 8.4 Types of coiled cables.

thermal damage is rare because data acquisition time is relatively short (i.e., less than 30 sec) using temporary mounting techniques.

Cables

Most portable vibration data collectors use a coiled cable to connect to the transducer (see Figure 8.4). The cable, much like a telephone cord, provides a relatively compact length when relaxed, but will extend to reach distant measurement points. For general use, this type of cable is acceptable, but it cannot be used for all applications.

The coiled cable is not acceptable for low-speed (i.e., less than 300 rpm) applications or where there is a strong electromagnetic field. Because of its natural tendency to return to its relaxed length, the coiled cable generates a low-level frequency that corresponds to the oscillation rate of the cable. In low-speed applications, this oscillation frequency can mask real vibration that is generated by the machine.

A strong electromagnetic field, such as that generated by large mill motors, accelerates cable oscillation. In these instances, the vibration generated by the cable will mask real machine vibration.

In applications where the coiled cable distorts or interferes with the accuracy of acquired data, a shielded coaxial cable should be used. Although these noncoiled cables can be more difficult to use in conjunction with a portable analyzer, they are absolutely essential for low-speed and electromagnetic field applications.

DATA MEASUREMENTS

Most vibration monitoring programs rely on data acquired from the machine housing or bearing caps. The only exceptions are applications that require direct measurement of actual shaft displacement to obtain an accurate picture of the machine's dynamics. This section discusses the number and orientation of measurement points required to profile a machine's vibration characteristics.

The fact that both normal and abnormal machine dynamics tend to generate unbalanced forces in one or more directions increases the analyst's ability to determine the root-cause of deviations in the machine's operating condition. Because of this, measurements should be taken in both radial and axial orientations.

Radial Orientation

Radially oriented measurements permit the analyst to understand the relationship of vibration levels generated by machine components where the forces are perpendicular to the shaft's centerline.

For example, mechanical imbalance generates radial forces in all directions, but misalignment generally results in a radial force in a single direction that corresponds with the misaligned direction. The ability to determine the actual displacement direction of the machine's shaft and other components greatly improves diagnostic accuracy.

Two radial measurement points located 90 degrees apart are required at each bearing cap. The two points permit the analyst to calculate the actual direction and relative amplitude of any displacement that is present within the machine.

Figure 8.5 illustrates a simple vector analysis where the vertical and horizontal radial readings acquired from the outboard bearing cap indicate a relative vertical vibration velocity of 0.5 inches per second peak (IPS-PK) and a horizontal vibration velocity of 0.3 IPS-PK. Using simple geometry, the amplitude of vibration velocity (0.583 IPS-PK) in the actual direction of deflection can be calculated.

Axial Orientation

Axially oriented measurements are used to determine the lateral movement of a machine's shaft or dynamic mass. These measurement points are oriented in-line or parallel with the shaft or direction of movement.

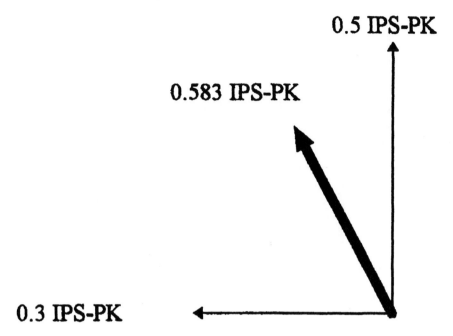

Figure 8.5 Resultant shaft velocity vector based on radial vibration measurements.

At least one axial measurement is required for each shaft or dynamic movement. In the case of shafts with a combination of float and fixed bearings, readings should be taken from the fixed or stationary bearing to obtain the best data.

TRANSDUCER-MOUNTING TECHNIQUES

For accuracy of data, a direct mechanical link between the transducer and the machine's casing or bearing cap is absolutely necessary. This makes the method used to mount the transducer crucial to obtaining accurate data. Slight deviations in this link will induce errors in the amplitude of vibration measurement and also may create false frequency components that have nothing to do with the machine.

Permanent Mounting

The best method of ensuring that the point of measurement, its orientation, and the compressive load are exactly the same each time is to permanently or hard mount the transducers, which is illustrated in Figure 8.6. This guarantees accuracy and repeatability of acquired data. However, it also increases the initial cost of the program. The average cost of installing a general-purpose accelerometer is about $300 per measurement point or $3000 for a typical machine-train.

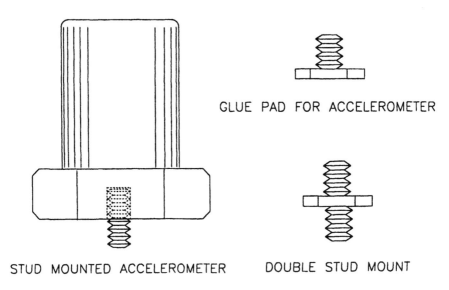

STUD MOUNTED ACCELEROMETER   DOUBLE STUD MOUNT

GLUE PAD FOR ACCELEROMETER

Figure 8.6 Permanent mounts provide best repeatability.

Quick-Disconnect Mounts

To eliminate the capital cost associated with permanently mounting transducers, a well-designed quick-disconnect mounting can be used instead. With this technique, a quick-disconnect stud having an average cost of less than $5 is permanently mounted at each measurement point. A mating sleeve built into the transducer is used to connect with the stud. A well-designed quick-disconnect mounting technique provides almost the same accuracy and repeatability as the permanent mounting technique, but at a much lower cost.

Magnets

For general-purpose use below 1000 Hz, a transducer can be attached to a machine by a magnetic base. Even though the resonant frequency of the transducer/magnet assembly may distort the data, this technique can be used with some success. However, since the magnet can be placed anywhere on the machine, it is difficult to guarantee that the exact location and orientation are maintained with each measurement. Figure 8.7 shows common magnetic mounts for transducers.

Handheld Transducer

Another method used by some plants to acquire data is handheld transducers. This approach is not recommended if it is possible to use any other method. Handheld transducers do not provide the accuracy and repeatability required to gain maximum benefit from a predictive maintenance program. If this technique must be used, extreme care

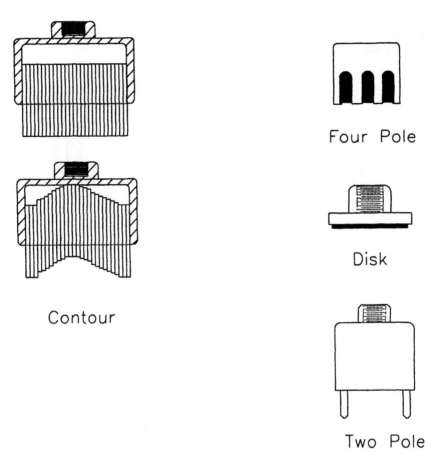

Four Pole

Disk

Contour

Two Pole

Figure 8.7 Common magnetic mounts for transducers.

should be exercised to ensure that the same location, orientation, and compressive load are used for every measurement. Figure 8.8 illustrates a handheld device.

ACQUIRING DATA

Three factors must be considered when acquiring vibration data: settling time, data verification, and additional data that may be required.

Settling Time

All vibration transducers require a power source that is used to convert mechanical motion or force to an electronic signal. In microprocessor-based analyzers, this power source is usually internal to the analyzer. When displacement probes are used, an external power source must be provided.

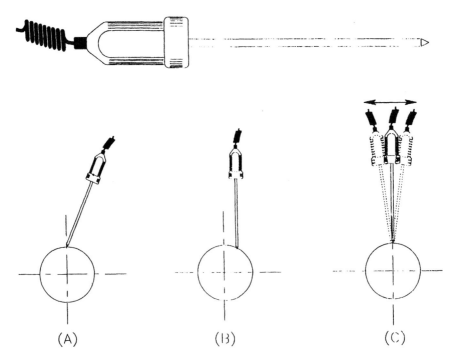

Figure 8.8 Handheld transducers should be avoided when possible. (a) Orientation is not 90 degrees to shaft centerline. (b) Measurement-point location is not always consistent. (c) Compressive load varies and may induce faulty readings.

When the power source is turned on, there is a momentary surge of power into the transducer. This surge distorts the vibration profile generated by the machine. Therefore, the data-acquisition sequence must include a time delay between powering up and acquiring data. The time delay will vary based on the specific transducer used and type of power source.

Some vibration analyzers include a user-selected time delay that can automatically be downloaded as part of the measurement route. If this feature is included, the delay can be preprogrammed for the specific transducer that will be used to acquire data. No further adjustment is required until the transducer type is changed.

In addition to the momentary surge created by energizing the power source, the mechanical action of placing the transducer on the machine creates a spike of energy that may distort the vibration profile. Therefore, the actual data-acquisition sequence should include a 10- to 20-second delay to permit decay of the spike created by mounting the transducer.

Data Verification

A number of equipment problems can result in bad or distorted data. In addition to the surge and spike discussed in the preceding section, damaged cables, transducers, power supplies, and other equipment failures can cause serious problems. Therefore, it is essential to verify all data throughout the acquisition process.

Most of the microprocessor-based vibration analyzers include features that facilitate verification of acquired data. For example, many include a low-level alert that automatically alerts the technician when acquired vibration levels are below a preselected limit. If these limits are properly set, the alert should be sufficient to detect this form of bad data.

Unfortunately, not all distortions of acquired data result in a low-level alert. Damaged or defective cables or transducers can result in a high level of low-frequency vibration. As a result, the low-level alert will not detect this form of bad data. However, the vibration signature will clearly display the abnormal profile that is associated with these problems.

In most cases, a defective cable or transducer generates a signature that contains a ski-slope profile, which begins at the lowest visible frequency and drops rapidly to the noise floor of the signature. If this profile is generated by defective components, it will not contain any of the normal rotational frequencies generated by the machine-train.

With the exception of mechanical rub, defective cables and transducers are the only sources of this ski-slope profile. When mechanical rub is present, the ski slope will also contain the normal rotational frequencies generated by the machine-train. In some cases, it is necessary to turn off the auto-scale function in order to see the rotational frequencies, but they will be clearly evident. If no rotational components are present, the cable and transducer should be replaced.

Additional Data

Data obtained from a vibration analyzer are not the only things required to evaluate machine-train or system condition. Variables, such as load, have a direct effect on the vibration profile of machinery and must be considered. Therefore, additional data should be acquired to augment the vibration profiles.

Most microprocessor-based vibration analyzers are capable of directly acquiring process variables and other inputs. The software and firmware provided with these systems generally support preprogrammed routes that include almost any direct or manual data input. These routes should include all data required to analyze effectively the operating condition of each machine-train and its process system.

Chapter 9

ANALYSIS TECHNIQUES

Techniques used in vibration analysis are trending, both broadband and narrowband; comparative analysis; and signature analysis.

TRENDING

Most vibration monitoring programs rely heavily on historical vibration-level amplitude trends as their dominant analysis tool. This is a valid approach if the vibration data are normalized to remove the influence of variables, such as load, on the recorded vibration energy levels. Valid trend data provide an indication of change over time within the monitored machine. As stated in preceding sections, a change in vibration amplitude is an indication of a corresponding change in operating condition that can be a useful diagnostic tool.

Broadband

Broadband analysis techniques have been used for monitoring the overall mechanical condition of machinery for more than 20 years. The technique is based on the overall vibration or energy from a frequency range of zero to the user-selected maximum frequency, F_{MAX}. Broadband data are overall vibration measurements expressed in units such as velocity-PK, acceleration-RMS, etc. This type of data, however, does not provide any indication of the specific frequency components that make up the machine's vibration signature. As a result, specific machine-train problems cannot be isolated and identified.

The only useful function of broadband analysis is long-term trending of the gross overall condition of machinery. Typically, a set of alert/alarm limits is established to monitor the overall condition of the machine-trains in a predictive maintenance program. However, this approach has limited value and, when used exclusively, severely limits the ability to achieve the full benefit of a comprehensive program.

Narrowband

Like broadband analysis, narrowband analysis also monitors the overall energy, but for a user-selected band of frequency components. The ability to select specific groups of frequencies, or narrowbands, increases the usefulness of the data. Using this technique can drastically reduce the manpower required to monitor machine-trains and improve the accuracy of detecting incipient problems.

Unlike broadband data, narrowband data provide the ability to directly monitor, trend, and alarm specific machine-train components automatically by the use of a microprocessor for a window of frequencies unique to specific machine components. For example, a narrowband window can be established to directly monitor the energy of a gear set that consists of the primary gear mesh frequency and corresponding side bands.

COMPARATIVE ANALYSIS

Comparative analysis directly compares two or more data sets in order to detect changes in the operating condition of mechanical or process systems. This type of analysis is limited to the direct comparison of the time-domain or frequency-domain signature generated by a machine. The method does not determine the actual dynamics of the system. Typically, the following data are used for this purpose: (1) baseline data, (2) known machine condition, or (3) industrial reference data.

Note that great care must be taken when comparing machinery vibration data to industry standards or baseline data. The analyst must make sure the frequency and amplitude are expressed in units and running speeds that are consistent with the standard or baseline data. The use of a microprocessor-based system with software that automatically converts and displays the desired terms offers a solution to this problem.

Baseline Data

Reference or baseline data sets should be acquired for each machine-train or process system to be included in a predictive maintenance program when the machine is installed or after the first scheduled maintenance once the program is established. These data sets can be used as a reference or comparison data set for all future measurements. However, such data sets must be representative of the normal operating condition of each machine-train. Three criteria are critical to the proper use of baseline comparisons: reset after maintenance, proper identification, and process envelope.

Reset After Maintenance

The baseline data set must be updated each time the machine is repaired, rebuilt, or when any major maintenance is performed. Even when best practices are used, machinery cannot be restored to as-new condition when major maintenance is performed. Therefore, a new baseline or reference data set must be established following these events.

Proper Identification

Each reference or baseline data set must be clearly and completely identified. Most vibration-monitoring systems permit the addition of a label or unique identifier to any user-selected data set. This capability should be used to clearly identify each baseline data set.

In addition, the data-set label should include all information that defines the data set. For example, any rework or repairs made to the machine should be identified. If a new baseline data set is selected after the replacement of a rotating element, this information should be included in the descriptive label.

Process Envelope

Because variations in process variables, such as load, have a direct effect on the vibration energy and the resulting signature generated by a machine-train, the actual operating envelope for each baseline data set must also be clearly identified. If this step is omitted, direct comparison of other data to the baseline will be meaningless. The label feature in most vibration monitoring systems permits tagging of the baseline data set with this additional information.

Known Machine Condition

Most microprocessor-based analyzers permit direct comparison to two machine-trains or components. The form of direct comparison, called cross-machine comparison, can be used to identify some types of failure modes.

When using this type of comparative analysis, the analyst compares the vibration energy and profile from a suspect machine to that of a machine with known operating condition. For example, the suspect machine can be compared to the baseline reference taken from a similar machine within the plant. Or, a machine profile with a known defect, such as a defective gear, can be used as a reference to determine if the suspect machine has a similar profile and, therefore, a similar problem.

Industrial Reference Data

One form of comparative analysis is direct comparison of the acquired data to industrial standards or reference values. The vibration-severity standards presented in Table

Table 9.1 Vibration Severity Standards*

Condition	Machine Classes			
	I	II	III	IV
Good operating condition	0.028	0.042	0.100	0.156
Alert limit	0.010	0.156	0.255	0.396
Alarm limit	0.156	0.396	0.396	0.622
Absolute fault limit	0.260	0.400	0.620	1.000

* Measurements are in inches per second peak. Applicable to a machine with running speed between 600 and 12,000 rpm. Narrowband setting: 0.3× to 3.0× running speed.

Machine Class Descriptions:

Class I	Small machine-trains or individual components integrally connected with the complete machine in its normal operating condition (i.e., drivers up to 20 hp).
Class II	Medium-sized machines (i.e., 20- to 100-hp drivers and 400-hp drivers on special foundations).
Class III	Large prime movers (i.e., drivers greater than 100 hp) mounted on heavy, rigid foundations.
Class IV	Large prime movers (i.e., drivers greater than 100 hp) mounted on relatively soft, lightweight structures.

Source: Derived by Integrated Systems, Inc. from ISO Standard #2372.

9.1 were established by the International Standards Organization (ISO). These data are applicable for comparison with filtered narrowband data taken from machine-trains with true running speeds between 600 and 12,000 rpm. The values from the table include all vibration energy between a lower limit of 0.3× true running speed and an upper limit of 3.0×. For example, an 1800-rpm machine would have a filtered narrowband between 540 (1800 × 0.3) and 5400 rpm (1800 × 3.0). A 3600-rpm machine would have a filtered narrowband between 1080 (3600 × 0.3) and 10,800 rpm (3600 × 3.0).

FAST FOURIER TRANSFORM SIGNATURE ANALYSIS

The phrase *full fast Fourier transform signature* is usually applied to the vibration spectrum that uniquely identifies a machine, component, system, or subsystem at a specific operating condition and time. It provides specific data on every frequency component within the overall frequency range of a machine-train. The typical frequency range can be from 0.1 to 20,000 Hz.

In microprocessor systems, the FFT signature is formed by breaking down the total frequency spectrum into unique components, or peaks. Each line or peak represents a specific frequency component that, in turn, represents one or more mechanical components within the machine-train. Typical microprocessor-based predictive maintenance systems can provide signature resolutions of at least 400 lines and many provide up to 12,800 lines.

A set of full-signature spectra can be an important analysis tool, but requires a tremendous amount of microprocessor memory. It is impractical to collect full, high-resolution spectra on all machine-trains on a routine basis. Data management and storage in the host computer is extremely difficult and costly. Full-range signatures should be collected only if a confirmed problem has been identified on a specific machine-train. This can be triggered automatically by exceeding a preset alarm limit in the historical amplitude trends.

Broadband and Full Signature

Systems that utilize either broadband or full signature measurements have limitations that may hamper the usefulness of the program. Broadband measurements usually do not have enough resolution at running speeds to be effective in early problem diagnostics. Full-signature measurement at every data point requires a massive data acquisition, handling, and storage system that greatly increases the capital and operating costs of the program.

Normally, a set of full-signature spectra is needed only when an identified machine-train problem demands further investigation. Please note that, although full signatures generate too much data for routine problem detection, they are essential for root-cause diagnostics. Therefore, the optimum system includes the capability to utilize all techniques. This ability optimizes the program's ability to trend, do full root-cause failure analysis, and still maintain minimum data management and storage requirements.

Narrowband

Typically, a machine-train's vibration signature is made up of vibration components with each component associated with one or more of the true running speeds within the machine-train. Because most machinery problems show up at or near one or more of the running speeds, the narrowband capability is very beneficial in that high-resolution windows can be preset to monitor the running speeds. However, many of the microprocessor-based predictive maintenance systems available do not have narrowband capability. Therefore, care should be taken to ensure that the system utilized does have this capability.

Part II

FREQUENCY-DOMAIN VIBRATION ANALYSIS

This module provides the basic knowledge and skills required to implement a computer-based vibration monitoring program. It discusses the following topics: (1) typical machine-train monitoring parameters, (2) database development, (3) data-acquisition equipment and methods, and (4) data analysis.

Although each of the commercially available computer-based vibration monitoring systems has unique features and formats, the information contained in this training module is applicable to all of the systems. However, the manual provided by the vendor should be used in conjunction with this module to ensure proper use of the microprocessor-based data collection analyzer and the computer-based software.

Chapter 10

OVERVIEW

During the past 10 years, vibration monitoring and analysis instrumentation have improved dramatically. During this period, a number of new microprocessor-based systems have been introduced that greatly simplify the collection, data management, long-term trending, and analysis of vibration data. While these advancements permit wider use of vibration monitoring as a predictive-maintenance tool, use is generally limited to relatively simple, steady-state rotating machinery. Typically, these systems collect single-channel, frequency-domain data.

Table 10.1 lists the typical machinery that can be monitored using these microprocessor-based systems. All machines included in the centrifugal column are ideal applications for this technology. Those in the columns headed Reciprocating and Machine-Trains can be evaluated, but the limitations of the technology preclude full diagnostic capability. The remaining machinery and process systems included under the heading Continuous Process are much more difficult to analyze using this technology, but it can be done.

Table 10.1 Typical Machinery Monitored by Vibration Analysis

Centrifugal	Reciprocating	Continuous Process
Pumps	Pumps	Continuous casters
Compressors	Compressors	Hot and cold strip lines
Blowers	Diesel engines	Annealing lines
Fans	Gasoline engines	Plating lines
Motor-generators	Cylinders and other machines	Paper machines
Ball mills		Can manufacturing lines
Chillers		Pickle lines
Product rolls		Printing

continued

Table 10.1 Typical Machinery Monitored by Vibration Analysis

Centrifugal	Machine-Trains	Continuous Process
Mixers	Grinders	Dyeing and finishing
Gearboxes	Boring machines	Roofing manufacturing lines
Centrifuges	Hobbing machines	Chemical production lines
Transmissions	Machining centers	Petroleum production lines
Turbines	Temper mills	Neoprene production lines
Generators	Metal-working machines	Polyester production lines
Rotary dryers	Rolling mills	Nylon production lines
Electric motors	Most machining equipment	Flooring production lines
All rotating machinery		Continuous process lines

ADVANTAGES

The automatic functions provided by most of the new systems have greatly reduced the time and manpower required to monitor critical plant equipment. These functions have virtually eliminated both the human errors and the setup time normally associated with older vibration-monitoring techniques.

Simplified Data Acquisition and Analysis

With the combined power of the data collector and system software, data acquisition has been reduced to simple measurement routes that require limited operator input. The technician's role is to temporarily mount a transducer at the proper measurement point and push a button. The microprocessor automatically acquires conditions, evaluates, and stores the vibration data.

Automated Data Management

Before computer-based systems were developed, a major limitation of vibration monitoring programs was the labor required to manage, store, retrieve, and analyze the massive amount of data generated. However, the computer-based systems in use today virtually eliminate this labor requirement. These systems automatically manage data and provide almost instant data retrieval for analysis.

LIMITATIONS

There are several limitations of the computer-based systems and some system characteristics, particularly simplified data acquisition and analysis, provide both advantages and disadvantages. Other limitations arise because only single-channel, steady-state, frequency-domain data greater than 600 cycles per minute (cpm) or 10 Hertz (Hz) can be collected. Note that cpm also is referred to as revolutions per minute (rpm).

Simplified Data Acquisition and Analysis

While providing many advantages, simplified data acquisition and analysis also can be a liability. If the database is improperly configured, the automated capabilities of these analyzers will yield faulty diagnostics that can allow catastrophic failure of critical plant machinery.

Because technician involvement is reduced to a minimum level, the normal tendency is to use untrained or partially trained personnel for this repetitive function. Unfortunately, the lack of training results in less awareness and knowledge of visual and audible clues that can, and should be, an integral part of the monitoring program.

Single-Channel Data

Most of the microprocessor-based vibration monitoring systems collect single-channel, steady-state data that cannot be used for all applications. Single-channel data are limited to the analysis of simple machinery that operates at relatively constant speed.

While most of the microprocessor-based instruments are limited to a single input channel, in some cases, a second channel is incorporated in the analyzer. However, this second channel generally is limited to input from a tachometer, or a once-per-revolution input signal. This second channel cannot be used for vibration-data capture.

This limitation prohibits the use of most microprocessor-based vibration analyzers for complex machinery or machines with variable speeds. Single-channel data-acquisition technology assumes that the vibration profile generated by a machine-train remains constant throughout the data-acquisition process. This is generally true in applications where machine speed remains relatively constant (i.e., within 5 to 10 rpm). In this case, its use does not severely limit diagnostic accuracy and can be effectively used in a predictive maintenance program.

Steady-State Data

Most of the microprocessor-based instruments are designed to handle steady-state vibration data. Few have the ability to reliably capture transient events such as rapid speed or load changes. As a result, their use is limited in situations where these occur.

In addition, vibration data collected with a microprocessor-based analyzer is filtered and conditioned to eliminate nonrecurring events and their associated vibration profiles. Antialiasing filters are incorporated into the analyzers specifically to remove spurious signals such as impacts. While the intent behind the use of antialiasing filters is valid, however, their use can distort a machine's vibration profile.

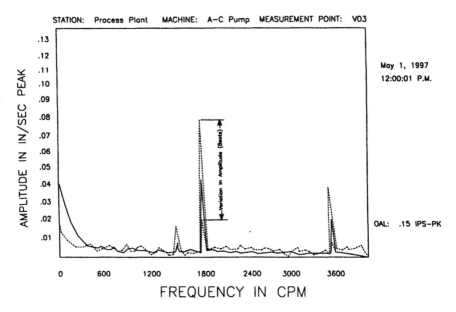

Figure 10.1 Vibration is dynamic and amplitudes constantly change.

Because vibration data are dynamic and the amplitudes constantly change as shown in Figure 10.1, most predictive maintenance system vendors strongly recommend averaging the data. They typically recommend acquiring 3 to 12 samples of the vibration profile and averaging the individual profiles into a composite signature. This approach eliminates the variation in vibration amplitude of the individual frequency components that make up the machine's signature. However, these variations, referred to as beats, can be a valuable diagnostic tool. Unfortunately, they are not available from microprocessor-based instruments because of averaging and other system limitations.

Frequency-Domain Data

Most predictive maintenance programs rely almost exclusively on frequency-domain vibration data. The microprocessor-based analyzers gather time-domain data and automatically convert it using fast Fourier transform (FFT) to frequency-domain data. A frequency-domain signature shows the machine's individual frequency components, or peaks.

While frequency-domain data analysis is much easier to learn than time-domain data analysis, it does not provide the ability to isolate and identify all incipient problems within the machine or its installed system. Because of this, additional techniques (e.g., time-domain, multichannel, and real-time analysis) must be used in conjunction with frequency-domain data analysis to obtain a complete diagnostic picture.

Low-Frequency Response

Many of the microprocessor-based vibration monitoring analyzers cannot capture accurate data from low-speed machinery or machinery that generates low-frequency vibration. Specifically, some of the commercially available analyzers cannot be used where frequency components are below 600 cpm or 10 Hz.

Two major problems restricting the ability to acquire accurate vibration data at low frequencies are electronic noise and the response characteristics of the transducer. The electronic noise of the monitored machine and the "noise floor" of the electronics within the vibration analyzer tend to override the actual vibration components found in low-speed machinery.

Analyzers specially equipped to handle noise are required for most industrial applications. There are at least three commercially available microprocessor-based analyzers capable of acquiring data below 600 cpm. These systems use special filters and data-acquisition techniques to separate real vibration frequencies from electronic noise. In addition, transducers with the required low-frequency response must be used.

Chapter 11

MACHINE-TRAIN MONITORING PARAMETERS

This chapter discusses normal failure modes, monitoring techniques that can prevent premature failures, and the measurement points required for monitoring common machine-train components. Understanding the specific location and orientation of each measurement point is critical to diagnosing incipient problems.

The frequency-domain, or fast Fourier transform (FFT), signature acquired at each measurement point is an actual representation of the individual machine-train component's motion at that point on the machine. Without knowing the specific location and orientation, it is difficult—if not impossible—to correctly identify incipient problems. In simple terms, the FFT signature is a photograph of the mechanical motion of a machine-train in a specific direction and at a specific point and time.

The vibration-monitoring process requires a large quantity of data to be collected, temporarily stored, and downloaded to a more powerful computer for permanent storage and analysis. In addition, there are many aspects to collecting meaningful data. Data collection generally is accomplished through the use of microprocessor-based data-collection equipment referred to as vibration analyzers. However, before analyzers can be used, it is necessary to set up a database with the data-collection and analysis parameters. The term *narrowband* refers to a specific frequency window that is monitored because of the knowledge that potential problems may occur due to known machine components or characteristics in this frequency range.

The orientation of each measurement point is an important consideration during the database setup and during analysis. There is an optimum orientation for each measurement point on every machine-train in a predictive maintenance program. For example, a helical gear set creates specific force vectors during normal operation. As the gear set degrades, these force vectors transmit the maximum vibration compo-

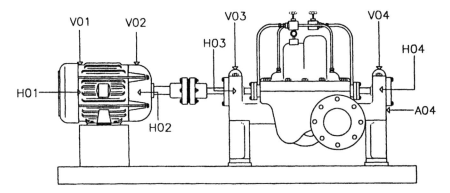

Figure 11.1 Recommended measurement-point logic: AO, axial orientation; HO, horizontal orientation; VO, vertical orientation.

nents. If only one radial reading is acquired for each bearing housing, it should be oriented in the plane that provides the greatest vibration amplitude.

For continuity, each machine-train should be set up on a "common-shaft" with the outboard driver bearing designated as the first data point. Measurement points should be numbered sequentially starting with the outboard driver bearing and ending with the outboard bearing of the final driven component. This is illustrated in Figure 11.1. Any numbering convention may be used, but it should be consistent, which provides two benefits:

1. Immediate identification of the location of a particular data point during the analysis/diagnostic phase.
2. Grouping the data points by "common shaft" enables the analyst to evaluate all parameters affecting each component of a machine-train.

DRIVERS

All machines require some form of motive power, which is referred to as a driver. This section includes the monitoring parameters for the two most common drivers: electric motors and steam turbines.

Electric Motors

Electric motors are the most common source of motive power for machine-trains. As a result, more of them are evaluated using microprocessor-based vibration monitoring systems than any other driver. The vibration frequencies of the following parameters are monitored to evaluate operating condition. This information is used to establish a database.

- Bearing frequencies
- Imbalance
- Line frequency
- Loose rotor bars
- Running speed
- Slip frequency
- V-belt intermediate drives.

Bearing Frequencies

Electric motors may incorporate either sleeve or rolling-element bearings. A narrow-band window should be established to monitor both the normal rotational and defect frequencies associated with the type of bearing used for each application.

Imbalance

Electric motors are susceptible to a variety of forcing functions that cause instability or imbalance. The narrowbands established to monitor the fundamental and other harmonics of actual running speed are useful in identifying mechanical imbalance, but other indices also should be used.

One such index is line frequency, which provides indications of instability. Modulations, or harmonics, of line frequency may indicate the motor's inability to find and hold magnetic center. Variations in line frequency also increase the amplitude of the fundamental and other harmonics of running speed.

Axial movement and the resulting presence of a third harmonic of running speed is another indication of instability or imbalance within the motor. The third harmonic is present whenever there is axial thrusting of a rotating element.

Line Frequency

Many electrical problems, or problems associated with the quality of the incoming power and internal to the motor, can be isolated by monitoring the line frequency. Line frequency refers to the frequency of the alternating current being supplied to the motor. In the case of 60-cycle power, monitoring of the fundamental or first harmonic (60 Hz), second harmonic (120 Hz), and third harmonic (180 Hz) should be performed.

Loose Rotor Bars

Loose rotor bars are a common failure mode of electric motors. Two methods can be used to identify them.

The first method uses high-frequency vibration components that result from oscillating rotor bars. Typically, these frequencies are well above the normal maximum frequency used to establish the broadband signature. If this is the case, a high-pass filter such as high-frequency domain can be used to monitor the condition of the rotor bars.

The second method uses the slip frequency to monitor for loose rotor bars. The passing frequency created by this failure mode energizes modulations associated with slip. This method is preferred since these frequency components are within the normal bandwidth used for vibration analysis.

Running Speed

The running speed of electric motors, both alternating current (AC) and direct current (DC), varies. Therefore, for monitoring purposes, these motors should be classified as variable-speed machines. A narrowband window should be established to track the true running speed.

Slip Frequency

Slip frequency is the difference between synchronous speed and actual running speed of the motor. A narrowband filter should be established to monitor electrical line frequency. The window should have enough resolution to clearly identify the frequency and the modulations, or sidebands, that represent slip frequency. Normally, these modulations are spaced at the difference between synchronous and actual speed, and the number of sidebands is equal to the number of poles in the motor.

V-Belt Intermediate Drives

Electric motors with V-belt intermediate drives display the same failure modes as those described previously. However, the unique V-belt frequencies should be monitored to determine if improper belt tension or misalignment is evident.

In addition, electric motors used with V-belt intermediate drive assemblies are susceptible to premature wear on the bearings. Typically, electric motors are not designed to compensate for the side loads associated with V-belt drives. In this type of application, special attention should be paid to monitoring motor bearings.

The primary data-measurement point on the inboard bearing housing should be located in the plane opposing the induced load (side load), with the secondary point at 90 degrees. The outboard primary data-measurement point should be in a plane opposite the inboard bearing with the secondary at 90 degrees.

Steam Turbines

There are wide variations in the size of steam turbines, which range from large utility units to small package units designed as drivers for pumps, etc. The following section describes in general terms the monitoring guidelines. Parameters that should be monitored are bearings, blade pass, mode shape (shaft deflection), and speed (both running and critical).

Bearings

Turbines use both rolling-element and Babbitt bearings. Narrowbands should be established to monitor both the normal rotational frequencies and failure modes of the specific bearings used in each turbine.

Blade Pass

Turbine rotors are comprised of a series of vanes or blades mounted on individual wheels. Each of the wheel units, which is referred to as a stage of compression, has a different number of blades. Narrowbands should be established to monitor the blade-pass frequency of each wheel. Loss of a blade or flexing of blades or wheels is detected by these narrowbands.

Mode Shape (Shaft Deflection)

Most turbines have relatively long bearing spans and highly flexible shafts. These factors, coupled with variations in process flow conditions, make turbine rotors highly susceptible to shaft deflection during normal operation. Typically, turbines operate in either the second or third mode and should have narrowbands at the second (2×) and third (3×) harmonics of shaft speed to monitor for mode shape.

Speed

All turbines are variable-speed drivers and operate near or above one of the rotor's critical speeds. Narrowbands should be established that track each of the critical speeds defined for the turbine's rotor. In most applications, steam turbines operate above the first critical speed and in some cases above the second. A movable narrowband window should be established to track the fundamental (1×), second (2×), and third (3×) harmonics of actual shaft speed. The best method is to use orders analysis and a tachometer to adjust the window location.

Normally, the critical speeds are determined by the mechanical design and should not change. However, changes in the rotor configuration or a buildup of calcium or other foreign materials on the rotor will affect them. The narrowbands should be wide enough to permit some increase or decrease.

INTERMEDIATE DRIVES

Intermediate drives transmit power from the primary driver to a driven unit or units. Included in this classification are chains, couplings, gearboxes, and V-belts.

Chains

In terms of its vibration characteristics, a chain-drive assembly is much like a gear set. The meshing of the sprocket teeth and chain links generates a vibration profile that is almost identical to that of a gear set. The major difference between these two machine-train components is that slack in the chain tends to modulate and amplify the tooth-mesh energy. Most of the forcing functions generated by a chain-drive assembly can be attributed to the forces generated by tooth-mesh. The typical frequencies associated with chain-drive assembly monitoring are those of running speed, tooth-mesh, and chain speed.

Running Speed

Chain drives are normally used to provide positive power transmission between a driver and driven unit where direct coupling cannot be accomplished. Chain drives generally have two distinct running speeds: driver or input speed and driven or output speed. Each of the shaft speeds is clearly visible in the vibration profile and a discrete narrowband window should be established to monitor each of the running speeds.

These speeds can be calculated using the ratio of the drive to driven sprocket. For example, where the drive sprocket has a circumference of 10 in. and the driven sprocket a circumference of 5 in., the output speed will be two times the input speed. Tooth-mesh narrowband windows should be created for both the drive and driven tooth-meshing frequencies. The windows should be broad enough to capture the sidebands or modulations that this type of passing frequency generates. The frequency of the sprocket-teeth meshing with the chain links, or passing frequency, is calculated by the following formula:

$$\text{Tooth-mesh Frequency} = \text{Number of Sprocket Teeth} \times \text{Shaft Speed}$$

Unlike gear sets, there can be two distinctive tooth-mesh frequencies for a chain-drive system. Because the drive and driven sprockets do not directly mesh, the meshing frequency generated by each sprocket is visible in the vibration profile.

Chain Speed

The chain acts much like a driven gear and has a speed that is unique to its length. The chain speed is calculated by the following equation:

$$\text{Chain Speed} = \frac{\text{Number of Drive Sprocket Teeth} \times \text{Shaft Speed}}{\text{Number of Links in Chain}}$$

For example:

$$\text{Chain Speed} = \frac{25 \text{ teeth} \times 100 \text{ rpm}}{250 \text{ links}} = \frac{2500}{250} = 10 \text{ cpm} = 10 \text{ rpm}$$

Couplings

Couplings cannot be monitored directly, but they generate forcing functions that affect the vibration profile of both the driver and driven machine-train component. Each coupling should be evaluated to determine the specific mechanical forces and failure modes they generate. This section discusses flexible couplings, gear couplings, jackshafts, and universal joints.

Flexible Couplings

Most flexible couplings use an elastomer or spring-steel device to provide power transmission from the driver to the driven unit. Both coupling types create unique mechanical forces that directly affect the dynamics and vibration profile of the machine-train.

The most obvious force with flexible couplings is endplay or movement in the axial plane. Both the elastomer and spring-steel devices have memory, which forces the axial position of both the drive and driven shafts to a neutral position. Because of their flexibility, these devices cause the shaft to move constantly in the axial plane. This is exhibited as harmonics of shaft speed. In most cases, the resultant profile is a signature that contains the fundamental (1x) frequency and second (2x) and third (3x) harmonics.

Gear Couplings

When properly installed and maintained, gear-type couplings do not generate a unique forcing function or vibration profile. However, excessive wear, variations in speed or torque, or overlubrication results in a forcing function.

Excessive wear or speed variation generates a gear-mesh profile that corresponds to the number of teeth in the gear coupling multiplied by the rotational speed of the driver. Since these couplings use a mating gear to provide power transmission, variations in speed or excessive clearance permit excitation of the gear-mesh profile.

Jackshafts

Some machine-trains use an extended or spacer shaft, called a jackshaft, to connect the driver and a driven unit. This type of shaft may use any combination of flexible coupling, universal joint, or splined coupling to provide the flexibility required making the connection. Typically, this type of intermediate drive is used either to absorb torsional variations during speed changes or to accommodate misalignment between the two machine-train components.

Because of the length of these shafts and the flexible couplings or joints used to transmit torsional power, jackshafts tend to flex during normal operation. Flexing results in a unique vibration profile that defines its operating mode shape.

In relatively low-speed applications, the shaft tends to operate in the first mode or with a bow between the two joints. This mode of operation generates an elevated vibration frequency at the fundamental (1x) turning speed of the jackshaft. In higher speed applications, or where the flexibility of the jackshaft increases, it deflects into an "S" shape between the two joints. This "S" or second mode shape generates an elevated frequency at both the fundamental (1x) frequency and the second harmonic (2x) of turning speed. In extreme cases, the jackshaft deflects further and operates in the

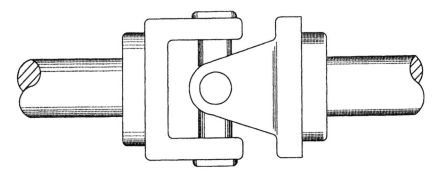

Figure 11.2 Typical double-pivot universal joint.

third mode. When this happens, it generates distinct frequencies at the fundamental (1×), second harmonic (2×), and third harmonic (3×) of turning speed.

As a general rule, narrowband windows should be established to monitor at least these three distinct frequencies, that is, 1×, 2×, and 3×. In addition, narrowbands should be established to monitor the discrete frequencies generated by the couplings or joints used to connect the jackshaft to the driver and driven unit.

Universal Joints

Various types of universal joints are used to transmit torsional power. In most cases, this type of intermediate drive is used where some misalignment between the drive and driven unit is necessary. Because of the misalignment, the universal's pivot points generate a unique forcing function that influences both the dynamics and vibration profile generated by a machine-train.

Figure 11.2 illustrates a typical double-pivot universal joint. This type of joint, which is similar to those used in automobiles, generates a unique frequency at four times (4×) the rotational speed of the shaft. Each of the pivot-point bearings generates a passing frequency each time the shaft completes a revolution.

Gearboxes

Gear sets are used to change speed or rotating direction of the primary driver. The basic monitoring parameters for all gearboxes include bearings, gear-mesh frequencies, and running speeds.

Bearings

A variety of bearing types is used in gearboxes. Narrowband windows should be established to monitor the rotational and defect frequencies generated by the specific type of bearing used in each application.

Special attention should be given to the thrust bearings, which are used in conjunction with helical gears. Because helical gears generate a relatively strong axial force, each gear shaft must have a thrust bearing located on the backside of the gear to absorb the thrust load. Therefore, all helical gear sets should be monitored for shaft runout.

The thrust, or positioning, bearing of a herringbone or double-helical gear has little or no normal axial loading. However, a coupling lockup can cause severe damage to the thrust bearing. Double-helical gears usually have only one thrust bearing, typically on the bull gear. Therefore, the thrust-bearing rotor should be monitored with at least one axial data-measurement point.

The primary data-measurement point on each shaft should be in a plane opposing the preload created by the gear mesh. A secondary data-measurement point should be located at 90 degrees to the primary point.

Gear-Mesh Frequencies

Each gear set generates a unique profile of frequency components that should be monitored. The fundamental gear-mesh frequency is equal to the number of teeth in the pinion or drive gear multiplied by the rotational shaft speed. In addition, each gear set generates a series of modulations, or sidebands, that surround the fundamental gear-mesh frequency. In a normal gear set, these modulations are spaced at the same frequency as the rotational shaft speed and appear on both sides of the fundamental gear mesh.

A narrowband window should be established to monitor the fundamental gear-mesh profile. The lower and upper limits of the narrowband should include the modulations generated by the gear set. The number of sidebands will vary depending on the resolution used to acquire data. In most cases, the narrowband limits should be about 10% above and below the fundamental gear-mesh frequency.

A second narrowband window should be established to monitor the second harmonic (2×) of gear mesh. Gear misalignment and abnormal meshing of gear sets result in multiple harmonics of the fundamental gear-mesh profile. This second window provides the ability to detect potential alignment or wear problems in the gear set.

Running Speeds

A narrowband window should be established to monitor each of the running speeds generated by the gear sets within the gearbox. The actual number of running speeds varies depending on the number of gear sets. For example, a single-reduction gearbox has two speeds: input and output. A double-reduction gearbox has three speeds: input, intermediate, and output. Intermediate and output speeds are determined by calculations based on input speed and the ratio of each gear set. Figure 11.3 illustrates a typical double-reduction gearbox.

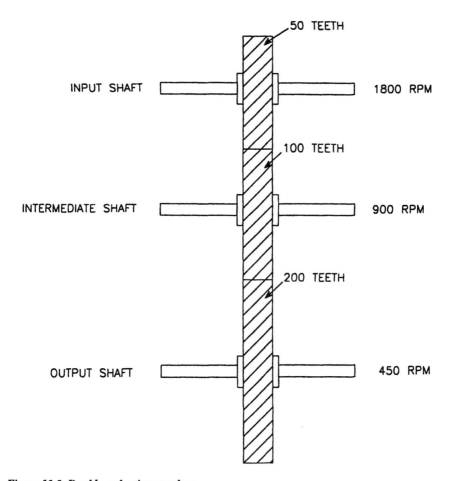

Figure 11.3 Double-reduction gearbox.

If the input speed is 1800 rpm, the intermediate and output speeds are calculated using the following equation:

$$\text{Intermediate Speed} = \frac{\text{Input Speed} \times \text{Number of Input Gear Teeth}}{\text{Number of Intermediate Gear Teeth}}$$

$$\text{Output Speed} = \frac{\text{Intermediate Speed} \times \text{Number of Intermediate Gear Teeth}}{\text{Number of Output Gear Teeth}}$$

V-Belts

V-belts are common intermediate drives for fans, blowers, and other types of machinery. Unlike some other power transmission mechanisms, V-belts generate unique forcing functions that must be understood and evaluated as part of a vibration analysis.

Table 11.1 Belt Drive Failure: Symptoms, Causes, and Corrective Actions

Symptom	Cause	Corrective Action
High 1× rotational frequency in radial direction.	Unbalanced or eccentric sheave.	Balance or replace sheave.
High 1× belt frequency with harmonics. Impacting at belt frequency in waveform.	Defects in belt.	Replace belt.
High 1× belt frequency. Sinusoidal waveform with period of belt frequency.	Unbalanced belt.	Replace belt.
High 1× rotational frequency in axial plane. 1× and possibly 2× radial.	Loose, misaligned, or mismatched belts.	Align sheaves; retension or replace belts as needed.

Source: Integrated Systems, Inc.

The key monitoring parameters for V-belt-driven machinery are fault frequency and running speed.

Most of the forcing functions generated by V-belt drives can be attributed to the elastic or rubber band effect of the belt material. This elasticity is needed to provide the traction required to transmit power from the drive sheave (i.e., pulley) to the driven sheave. Elasticity causes belts to act like springs, increasing vibration in the direction of belt wrap, but damping it in the opposite direction. As a result, belt elasticity tends to accelerate wear and the failure rate of both the driver and driven unit.

Fault Frequencies

Belt-drive fault frequencies are the frequencies of the driver, the driven unit, and the belt. In particular, frequencies at 1× the respective shaft speeds indicate faults with the balance, concentricity, and alignment of the sheaves. The belt frequency and its harmonics indicate problems with the belt. Table 11.1 summarizes the symptoms and causes of belt-drive failures, as well as corrective actions.

Running Speeds

Belt-drive ratios may be calculated if the pitch diameters (see Figure 11.4) of the sheaves are known. This coefficient, which is used to determine the driven speed given the drive speed, is obtained by dividing the pitch diameter of the drive sheave by the pitch diameter of the driven sheave. These relationships are expressed by the following equations:

$$\text{Drive Reduction} = \frac{\text{Drive Sheave Diameter}}{\text{Driven Sheave Diameter}}$$

$$\text{Driven Speed, rpm} = \text{Drive Speed, rpm} \times \left(\frac{\text{Drive Sheave Diameter}}{\text{Driven Sheave Diameter}}\right)$$

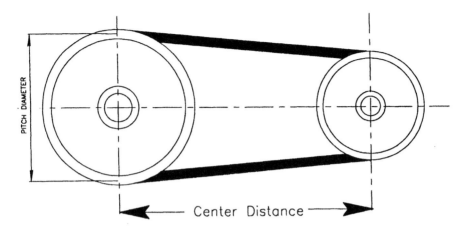

Figure 11.4 Pitch diameter and center-to-center distance between belt sheaves.

$$\text{Drive Speed, rpm} = \text{Driven Speed, rpm} \times \left(\frac{\text{Driven Sheave Diameter}}{\text{Drive Sheave Diameter}} \right)$$

Using these relationships, the sheave rotational speeds can be determined. However, obtaining the other component speeds requires a bit more effort. The rotational speed of the belt cannot be directly determined using the information presented so far. To calculate belt rotational speed (rpm), the linear belt speed must first be determined by finding the linear speed (in./min) of the sheave at its pitch diameter. In other words, multiply the pitch circumference (PC) by the rotational speed of the sheave, where:

$$\text{Pitch Circumference (in)} = \pi \times \text{Pitch Diameter (in)}$$

$$\text{Linear Speed (in/min)} = \text{Pitch Circumference (in)} \times \text{Sheave Speed (rpm)}$$

To find the exact rotational speed of the belt (rpm), divide the linear speed by the length of the belt:

$$\text{Belt Rotational Speed (rpm)} = \frac{\text{Linear Speed (in/min)}}{\text{Belt Length (in)}}$$

To approximate the rotational speed of the belt, the linear speed may be calculated using the pitch diameters and the center-to-center distance (see Figure 11.4) between the sheaves. This method is accurate only if there is no belt sag. Otherwise, the belt rotational speed obtained using this method is slightly higher than the actual value.

In the special case where the drive and driven sheaves have the same diameter, the formula for determining the belt length is as follows:

$$\text{Belt Length} = \text{Pitch Circumference} + (2 \times \text{Center Distance})$$

The following equation is used to approximate the belt length where the sheaves have different diameters:

$$\text{Belt Length} = \frac{\text{Drive PC} + \text{Driven PC}}{2} + (2 \times \text{Center Distance})$$

DRIVEN COMPONENTS

This chapter cannot effectively discuss all possible combinations of driven components that may be found in a plant. However, the guidelines provided here can be used to evaluate most of the machine-trains and process systems that are typically included in a microprocessor-based vibration monitoring program.

Compressors

The two basic types of compressors are (1) centrifugal and (2) positive displacement. Both of these major classifications can be further divided into subtypes, depending on their operating characteristics. This section provides an overview of the more common centrifugal and positive-displacement compressors.

Centrifugal

The two types of commonly used centrifugal compressors are (1) in-line and (2) bullgear.

In-Line
The in-line centrifugal compressor functions in exactly the same manner as a centrifugal pump. The only difference between the pump and the compressor is that the compressor has smaller clearances between the rotor and casing. Therefore, in-line centrifugal compressors should be monitored and evaluated in the same manner as centrifugal pumps and fans. As with these driven components, the in-line centrifugal compressor is comprised of a single shaft with one or more impeller(s) mounted on the shaft. All components generate simple rotating forces that can be monitored and evaluated with ease. Figure 11.5 shows a typical in-line centrifugal compressor.

Bullgear
The bullgear centrifugal compressor (Figure 11.6) is a multistage unit that utilizes a large helical gear mounted on the compressor's driven shaft and two or more pinion gears, which drive the impellers. These impellers act in series, whereby compressed air or gas from the first-stage impeller discharge is directed by flow channels within the compressor's housing to the second-stage inlet. The discharge of the second stage is channeled to the inlet of the third stage. This channeling occurs until the air or gas exits the final stage of the compressor.

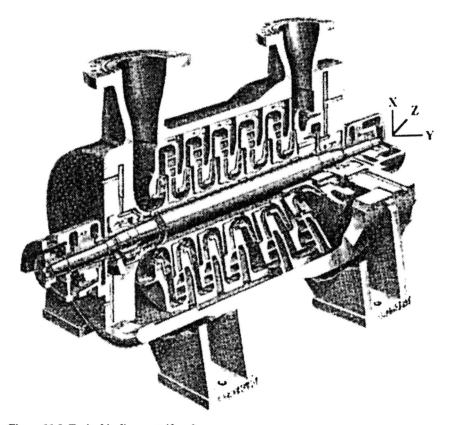

Figure 11.5 Typical in-line centrifugal compressor.

Generally, the driver and bullgear speed is 3600 rpm or less, and the pinion speeds are as high as 60,000 rpm (see Figure 11.7). These machines are produced as a package with the entire machine-train mounted on a common foundation that also includes a panel with control and monitoring instrumentation.

Positive Displacement

Positive-displacement compressors, also referred to as dynamic-type compressors, confine successive volumes of fluid within a closed space. The pressure of the fluid increases as the volume of the closed space decreases.

Reciprocating

Reciprocating compressors are positive-displacement types having one or more cylinders. Each cylinder is fitted with a piston driven by a crankshaft through a connecting rod. As the name implies, compressors within this classification displace a fixed volume of air or gas with each complete cycle of the compressor.

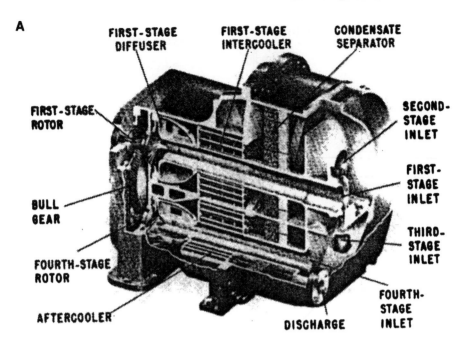

A

FIRST-STAGE DIFFUSER
FIRST-STAGE INTERCOOLER
CONDENSATE SEPARATOR
FIRST-STAGE ROTOR
SECOND-STAGE INLET
FIRST-STAGE INLET
BULL GEAR
THIRD-STAGE INLET
FOURTH-STAGE ROTOR
AFTERCOOLER
DISCHARGE
FOURTH-STAGE INLET

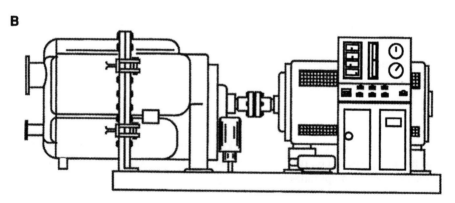

B

Figure 11.6 (a) Cut-away of bullgear centrifugal compressors (b) Bullgear centrifugal compressors have built-in supervisory systems.

Reciprocating compressors have unique operating dynamics that directly affect their vibration profiles. Unlike most centrifugal machinery, reciprocating machines combine rotating and linear motions that generate complex vibration signatures.

Crankshaft Frequencies
All reciprocating compressors have one or more crankshaft(s) that provide the motive power to a series of pistons, which are attached by piston arms. These crankshafts rotate in the same manner as the shaft in a centrifugal machine. However, their

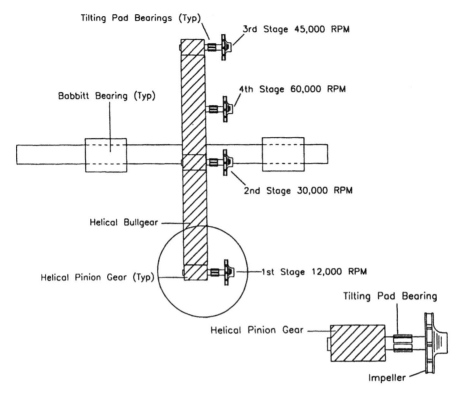

Figure 11.7 Internal bullgear drives' pinion gears at each stage.

dynamics are somewhat different. The crankshafts generate all of the normal frequencies of a rotating shaft (i.e., running speed, harmonics of running speed, and bearing frequencies), but the amplitudes are much higher.

In addition, the relationship of the fundamental (1×) frequency and its harmonics changes. In a normal rotating machine, the 1× frequency normally contains between 60 and 70% of the overall, or broadband, energy generated by the machine-train. In reciprocating machines, however, this profile changes. Two-cycle reciprocating machines, such as single-action compressors, generate a high second harmonic (2×) and multiples of the second harmonic. While the fundamental (1×) is clearly present, it is at a much lower level.

Frequency Shift Due to Pistons
The shift in vibration profile is the result of the linear motion of the pistons used to provide compression of the air or gas. As each piston moves through a complete

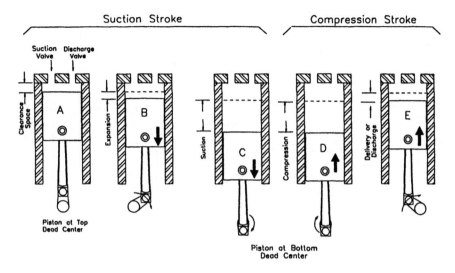

Figure 11.8 Two-cycle, or single-action, air compressor cylinder.

cycle, it must change direction two times. This reversal of direction generates the higher second harmonic (2×) frequency component.

In a two-cycle machine, all pistons complete a full cycle each time the crankshaft completes one revolution. Figure 11.8 illustrates the normal action of a two-cycle, or single-action, compressor. Inlet and discharge valves are located in the clearance space and connected through ports in the cylinder head to the inlet and discharge connections. During the suction stroke, the compressor piston starts its downward stroke and the air under pressure in the clearance space rapidly expands until the pressure falls below that on the opposite side of the inlet valve (point B). This difference in pressure causes the inlet valve to open into the cylinder until the piston reaches the bottom of its stroke (point C).

During the compression stroke, the piston starts upward, compression begins, and at point D has reached the same pressure as the compressor intake. The spring-loaded inlet valve then closes. As the piston continues upward, air is compressed until the pressure in the cylinder becomes great enough to open the discharge valve against the pressure of the valve springs and the pressure of the discharge line (point E). From this point, to the end of the stroke (point E to point A), the air compressed within the cylinder is discharged at practically constant pressure.

The impact energy generated by each piston as it changes direction is clearly visible in the vibration profile. Since all pistons complete a full cycle each time the crankshaft completes one full revolution, the total energy of all pistons is displayed at the fundamental (1×) and second harmonic (2×) locations.

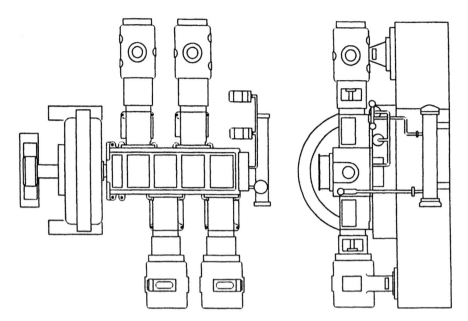

Figure 11.9 Horizontal reciprocating compressor.

In a four-cycle machine, two complete revolutions (720 degrees) are required for all cylinders to complete a full cycle.

Piston Orientations
Crankshafts on positive-displacement reciprocating compressors have offsets from the shaft centerline that provide the stroke length for each piston. The orientation of the offsets has a direct effect on the dynamics and vibration amplitudes of the compressor. In an opposed-piston compressor where pistons are 180 degrees apart, the impact forces as the pistons change directions are reduced. As one piston reaches top dead center, the opposing piston also is at top dead center. The impact forces, which are 180 degrees out of phase, tend to cancel or balance each other as the two pistons change directions.

Another configuration, called an unbalanced design, has piston orientations that are neither in phase nor 180 degrees out of phase. In these configurations, the impact forces generated as each piston changes direction are not balanced by an equal and opposite force. As a result, the impact energy and the vibration amplitude are greatly increased.

Horizontal reciprocating compressors (see Figure 11.9) should have *X-Y* data points on both the inboard and outboard main crankshaft bearings, if possible, to monitor the connecting rod or plunger frequencies and forces.

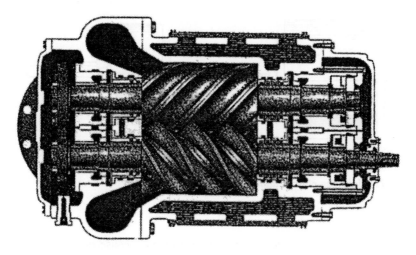

Figure 11.10 Screw compressors—steady-state applications only.

Screw

Screw compressors have two rotors with interlocking lobes and act as positive-displacement compressors (see Figure 11.10). This type of compressor is designed for baseload, or steady-state, operation and is subject to extreme instability should either the inlet or discharge conditions change. Two helical gears mounted on the outboard ends of the male and female shafts synchronize the two rotor lobes.

Analysis parameters should be established to monitor the key indices of the compressor's dynamics and failure modes. These indices should include bearings, gear mesh, rotor passing frequencies, and running speed. However, because of its sensitivity to process instability and the normal tendency to thrust, the most critical monitoring parameter is axial movement of the male and female rotors.

Bearings

Screw compressors use both Babbitt and rolling-element bearings. Because of the thrust created by process instability and the normal dynamics of the two rotors, all screw compressors use heavy-duty thrust bearings. In most cases, they are located on the outboard end of the two rotors, but some designs place them on the inboard end. The actual location of the thrust bearings must be known and used as a primary measurement-point location.

Gear Mesh

The helical timing gears generate a meshing frequency equal to the number of teeth on the male shaft multiplied by the actual shaft speed. A narrowband window should be created to monitor the actual gear mesh and its modulations. The limits of the window should be broad enough to compensate for a variation in speed between full load and no load.

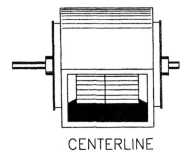
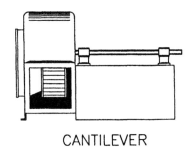

CENTERLINE   CANTILEVER

Figure 11.11 Major fan classifications.

The gear set should be monitored for axial thrusting. Because of the compressor's sensitivity to process instability, the gears are subjected to extreme variations in induced axial loading. Coupled with the helical gear's normal tendency to thrust, the change in axial vibration is an early indicator of incipient problems.

Rotor Passing
The male and female rotors act much like any bladed or gear unit. The number of lobes on the male rotor multiplied by the actual male shaft speed determines the rotor-passing frequency. In most cases, there are more lobes on the female than on the male. To ensure inclusion of all passing frequencies, the rotor-passing frequency of the female shaft also should be calculated. The passing frequency is equal to the number of lobes on the female rotor multiplied by the actual female shaft speed.

Running Speeds
The input, or male, rotor in screw compressors generally rotates at a no-load speed of either 1800 or 3600 rpm. The female, or driven, rotor operates at higher no-load speeds ranging between 3600 and 9000 rpm. Narrowband windows should be established to monitor the actual running speed of the male and female rotors. The windows should have an upper limit equal to the no-load design speed and a lower limit that captures the slowest, or fully loaded, speed. Generally, the lower limits are between 15 and 20% lower than no-load.

Fans

Fans have many different industrial applications and designs vary. However, all fans fall into two major categories: (1) centerline and (2) cantilever. The centerline configuration has the rotating element located at the midpoint between two rigidly supported bearings. The cantilever or overhung fan has the rotating element located outboard of two fixed bearings. Figure 11.11 illustrates the difference between the two fan classifications.

The following parameters are monitored in a typical predictive maintenance program for fans: aerodynamic instability, running speeds, and shaft mode shape, or shaft deflection.

Aerodynamic Instability

Fans are designed to operate in a relatively steady-state condition. The effective control range is typically 15 to 30% of their full range. Operation outside of the effective control range results in extreme turbulence within the fan, which causes a marked increase in vibration. In addition, turbulent flow caused by restricted inlet airflow, leaks, and a variety of other factors increases rotor instability and the overall vibration generated by a fan.

Both of these abnormal forcing functions (i.e., turbulent flow and operation outside of the effective control range) increase the level of vibration. However, when the instability is relatively minor, the resultant vibration occurs at the vane-pass frequency. As it becomes more severe, there also is a marked increase in the broadband energy.

A narrowband window should be created to monitor the vane-pass frequency of each fan. The vane-pass frequency is equal to the number of vanes or blades on the fan's rotor multiplied by the actual running speed of the shaft. The lower and upper limits of the narrowband should be set about 10% above and below ($\pm$10%) the calculated vane-pass frequency. This compensates for speed variations and it includes the broadband energy generated by instability.

Running Speeds

Fan running speed varies with load. If fixed filters are used to establish the bandwidth and narrowband windows, the running speed upper limit should be set to the synchronous speed of the motor, and the lower limit set at the full-load speed of the motor. This setting provides the full range of actual running speeds that should be observed in a routine monitoring program.

Shaft Mode Shape (Shaft Deflection)

The bearing-support structure is often inadequate for proper shaft support because of its span and stiffness. As a result, most fans tend to operate with a shaft that deflects from its true centerline. Typically, this deflection results in a vibration frequency at the second (2x) or third (3x) harmonic of shaft speed.

A narrowband window should be established to monitor the fundamental (1x), second (2x), and third (3x) harmonic of shaft speed. With these windows, the energy associated with shaft deflection, or mode shape, can be monitored.

Generators

As with electric motor rotors, generator rotors always seek the magnetic center of their casings. As a result, they tend to thrust in the axial direction. In almost all cases, this axial movement, or endplay, generates a vibration profile that includes the fundamental (1×), second (2×) and third (3×) harmonic of running speed. Key monitoring parameters for generators include bearings, casing and shaft, line frequency, and running speed.

Bearings

Large generators typically use Babbitt bearings, which are nonrotating, lined metal sleeves (also referred to as fluid-film bearings) that depend on a lubricating film to prevent wear. However, these bearings are subjected to abnormal wear each time a generator is shut off or started. In these situations, the entire weight of the rotating element rests directly on the lower half of the bearings. When the generator is started, the shaft climbs the Babbitt liner until gravity forces the shaft to drop to the bottom of the bearing. This alternating action of climb and fall is repeated until the shaft speed increases to the point that a fluid film is created between the shaft and Babbitt liner.

Subharmonic frequencies (i.e., less than the actual shaft speed) are the primary evaluation tool for fluid-film bearings and they must be monitored closely. A narrowband window that captures the full range of vibration frequency components between electronic noise and running speed is an absolute necessity.

Casing and Shaft

Most generators have relatively soft support structures. Therefore, they require shaft vibration monitoring measurement points in addition to standard casing measurement points. This requires the addition of permanently mounted proximity, or displacement, transducers that can measure actual shaft movement.

The third (3×) harmonic of running speed is a critical monitoring parameter. Most, if not all, generators tend to move in the axial plane as part of their normal dynamics. Increases in axial movement, which appear in the third harmonic, are early indicators of problems.

Line Frequency

Many electrical problems cause an increase in the amplitude of line frequency, typically 60 Hz, and its harmonics. Therefore, a narrowband should be established to monitor the 60-, 120-, and 180-Hz frequency components.

Running Speed

Actual running speed remains relatively constant on most generators. While load changes create slight variations in actual speed, the change in speed is minor. Generally, a narrowband window with lower and upper limits of ±10% of design speed is sufficient.

Process Rolls

Process rolls are commonly found in paper machines and other continuous process applications. Process rolls generate few unique vibration frequencies. In most cases, the only vibration frequencies generated are running speed and bearing rotational frequencies.

However, rolls are highly prone to loads induced by the process. In most cases, rolls carry some form of product or a mechanism that, in turn, carries a product. For example, a simple conveyor has rolls that carry a belt, which carries product from one location to another. The primary monitoring parameters for process rolls include bearings, load distribution, and misalignment.

Bearings

Both nonuniform loading and roll misalignment change the bearing load zones. In general, either of these failure modes results in an increase in outer-race loading. This is caused by the failure mode forcing the full load onto one quadrant of the bearing's outer race.

Therefore, the ball-pass outer-race frequency should be monitored closely on all process rolls. Any increase in this unique frequency is a prime indication of a load, tension, or misaligned roll problem.

Load Distribution

By design, process rolls should be uniformly loaded across their entire bearing span (see Figure 11.12). Improper tracking and/or tension of the belt, or product carried by the rolls, will change the loading characteristics.

The loads induced by the belt increase the pressure on the loaded bearing and decrease the pressure on the unloaded bearing. An evaluation of process rolls should include a cross-comparison of the overall vibration levels and the vibration signature of each roll's inboard and outboard bearing.

Misalignment

Misalignment of process rolls is a common problem. On a continuous process line, most rolls are mounted in several levels. The distance between the rolls and the change in elevation make it extremely difficult to maintain proper alignment.

In a vibration analysis, roll misalignment generates a signature similar to classical parallel misalignment. It generates dominant frequencies at the fundamental (1x) and second (2x) harmonic of running speed.

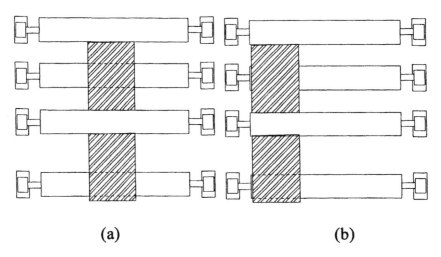

(a)   (b)

Figure 11.12 Rolls should be uniformly loaded: (a) proper and (b) improper.

Pumps

A wide variety of pumps are used by industry and they can be grouped into two types: centrifugal and positive displacement. Pumps are highly susceptible to process-induced or installation-induced loads. Some pump designs are more likely to have axial- or thrust-induced load problems. Induced loads created by hydraulic forces also are a serious problem in most pump applications.

Recommended monitoring for each type of pump is essentially the same, regardless of specific design or manufacturer. However, process variables such as flow, pressure, load, etc., must be taken into account.

Centrifugal

Centrifugal pumps can be divided into two basic types: end-suction and horizontal split-case. These two major classifications can be broken further into single-stage and multistage pumps. Each of these classifications has common monitoring parameters, but each also has unique features that alter their forcing functions and the resultant vibration profile. The common monitoring parameters for all centrifugal pumps include axial thrusting, vane-pass, and running speed.

Axial Thrusting
End-suction and multistage pumps with in-line impellers are prone to excessive axial thrusting. In the end-suction pump, the centerline axial inlet configuration is the primary source of thrust. Restrictions in the suction piping, or low suction pressures, create a strong imbalance that forces the rotating element toward the inlet.

Multistage pumps with in-line impellers generate a strong axial force on the outboard end of the pump. Most of these pumps have oversized thrust bearings (e.g., Kingsbury bearings) that restrict the amount of axial movement. However, bearing wear caused by constant rotor thrusting is a dominant failure mode. Monitoring of the axial movement of the shaft should be done whenever possible.

Hydraulic Instability (Vane Pass)

Hydraulic or flow instability is common in centrifugal pumps. In addition to the restrictions of the suction and discharge discussed previously, the piping configuration in many applications creates instability. Although flow through the pump should be laminar, sharp turns or other restrictions in the inlet piping can create turbulent flow conditions. Forcing functions such as these result in hydraulic instability, which displaces the rotating element within the pump.

In a vibration analysis, hydraulic instability is displayed at the vane-pass frequency of the pump's impeller. Vane-pass frequency is equal to the number of vanes in the impeller multiplied by the actual running speed of the shaft. Therefore, a narrowband window should be established to monitor the vane-pass frequency of all centrifugal pumps.

Running Speed

Most pumps are considered constant speed, but the true speed changes with variations in suction pressure and back-pressure caused by restrictions in the discharge piping. The narrowband should have lower and upper limits sufficient to compensate for these speed variations. Generally, the limits should be set at speeds equal to the full-load and no-load ratings of the driver.

There is a potential for unstable flow through pumps, which is created by both the design-flow pattern and the radial deflection caused by back-pressure in the discharge piping. Pumps tend to operate at their second-mode shape or deflection pattern. This mode of operation generates a unique vibration frequency at the second harmonic (2x) of running speed. In extreme cases, the shaft may be deflected further and operate in its third (3x) mode shape. Therefore, both of these frequencies should be monitored.

Positive Displacement

A variety of positive-displacement pumps are commonly used in industrial applications. Each type has unique characteristics that must be understood and monitored. However, most of the major types have common parameters that should be monitored.

With the exception of piston-type pumps, most of the common positive-displacement pumps utilize rotating elements to provide a constant-volume, constant-pressure output. As a result, these pumps can be monitored with the following parameters: hydraulic instability, passing frequencies, and running speed.

Hydraulic Instability (Vane Pass)

Positive-displacement pumps are subject to flow instability, which is created either by process restrictions or by the internal pumping process. Increases in amplitude at the passing frequencies, as well as harmonics of both shaft running speed and the passing frequencies, typically result from instability.

Passing Frequencies

With the exception of piston-type pumps, all positive-displacement pumps have one or more passing frequencies generated by the gears, lobes, vanes, or wobble plates used in different designs to increase the pressure of the pumped liquid. These passing frequencies can be calculated in the same manner as the blade or vane-passing frequencies in centrifugal pumps (i.e., multiplying the number of gears, lobes, vanes, or wobble plates times the actual running speed of the shaft).

Running Speeds

All positive-displacement pumps have one or more rotating shafts that provide power transmission from the primary driver. Narrowband windows should be established to monitor the actual shaft speeds, which are in most cases essentially constant. Upper and lower limits set at ±10% of the actual shaft speed are usually sufficient.

Chapter 12

DATABASE DEVELOPMENT

Valid data are an absolute prerequisite of vibration monitoring and analysis. Without accurate and complete data taken in the appropriate frequency range, it is impossible to interpret the vibration profiles obtained from a machine-train.

This is especially true in applications that use microprocessor/computer-based systems. These systems require a database that specifies the monitoring parameters, measurement routes, analysis parameters, and a variety of other information. This input is needed to acquire, trend, store, and report what is referred to as "conditioned" vibration data.

The steps in developing such a database are (1) collection of machine and process data and (2) database setup. Input requirements of the software are machine and process specifications, analysis parameters, data filters, alert/alarm limits, and a variety of other parameters used to automate the data-acquisition process.

MACHINE AND PROCESS DATA COLLECTION

Database development can be accelerated and its accuracy improved by first creating detailed equipment and process information sheets that fully describe each machine and system to be monitored.

Equipment Information Sheets

The first step in establishing a database that defines the operating condition of each machine-train or production system is to generate an equipment information sheet (EIS) for each machine-train. The information sheet must contain all of the machine-specific data such as type of operation and information on all of the components that make up the machine-train.

Type of Operation

The EIS should define the type of operation (i.e., constant speed or variable speed) that best describes the normal operation of each machine-train. This information allows the analyst to determine the best method of monitoring and evaluating each machine.

Constant-Speed Machinery

Few, if any, machines found in a manufacturing or production plant are truly constant speed. While the nameplate and specifications may indicate that a machine operates at a fixed speed, it will vary slightly in normal operation.

The reason for speed variations in constant-speed machinery is variation of process load. For example, a centrifugal pump's load will vary due to the viscosity of the fluid being pumped or changes in suction or discharge pressure. The pump speed will change as a result of these load changes.

As a general rule, the speed variation in a constant-speed machine is about 15%. For an electric motor, the actual variation can be determined by obtaining the difference between the amperage drawn under full-load and no-load conditions. This difference, taken as a percentage of full-load amperage draw, provides the actual percentage of speed-range variation that can be expected.

Variable-Speed Machinery

For machinery and process systems that have a wide range of operating speeds, the data sheet should provide a minimum and maximum speed that can be expected during normal operation. In addition, a complete description of other variables (e.g., product type) that affect the machine's speed should be included. For example, a process line may operate at 500 ft/min with product A and 1000 ft/min with product B. Therefore, the data sheet must define all of the variables associated with both product A and product B.

Constant Versus Variable Load

As with constant-speed machines, true constant-load machines are rare. For the few that may be specified as having constant load, there are factors that cause load changes to occur. These factors include variations in product, operating conditions, and ambient environment. These variations will have a direct, and often dramatic, impact on a machine's vibration profile.

Variations in load, no matter how slight, alter the vibration profile generated by a machine or system. The relationship between load and the vibration energy generated by a machine can be a multiple of four. In other words, a 10% change in load may increase or decrease the vibration energy by 40%.

When using vibration data as a diagnostic tool, you must always adjust or normalize the data to the actual load that was present when the data set was acquired.

Machine Components

The EIS should provide information on all components (e.g., bearings, gears, gear-boxes, electric motors, pumps) that make up the machine-train. Because these components generate vibration energy and unique frequency components, this information is essential for proper analysis. At a minimum, the information sheet must include detailed bearing information, passing frequencies, and nameplate data.

Bearings

All bearings in a machine-train must be identified. For example, rolling-element bearing data must include the manufacturer's part number, bearing geometry, and the unique rotational frequencies that it will generate. Rotational frequencies can be determined by using the bearing part number to look them up in the database that is included in most vibration monitoring software programs. They also can be obtained from the bearing vendor. Babbitt or sleeve bearing data must include type (e.g., plain, tilting-pad), as well as manufacturer and part number.

Passing Frequencies

All components that generate a passing frequency must be included on the information sheet. Such components include fan or compressor blades, vanes on pump impellers, rotor bars in electric motors, and gear teeth on both the pinion and bullgear in a gear set. The number of vanes, blades, and gear teeth must be recorded on the information sheet. The passing frequency is the number of vanes, blades, etc., times the rotation speed of the shaft on which they are mounted.

Nameplate Data

Each machine-train component has a vendor's nameplate permanently attached to its housing. The EIS should include all nameplate data, including the serial number that uniquely identifies a machine or component.

Knowing the serial number allows detailed information on a machine or component to be obtained. This is possible because machinery manufacturers must maintain records for their products. These records, which are usually identified by serial number, contain complete design and performance data for that specific unit. For example, it is possible to obtain a performance curve or complete bill of materials for each pump found in a plant.

Process Information Sheets

A process information sheet (PIS) should be developed for each machine-train and production process that is to be included in a predictive maintenance program. These data sheets should include all process variables that affect the dynamics and vibration profiles of the monitored components.

Many production and process systems handle a wide range of products. They typically have radically different machine and system operating parameters, as well as variable speeds and loads for each of the products they process.

Each process parameter directly affects both the machinery dynamics and the vibration profiles. For example, the line tension, strip width, and hardness of the incoming strip radically affect the vibration profile generated by a continuous process line in a steel mill. With few exceptions, process variations such as these must be considered in the vibration analysis.

DATABASE SETUP

The input-data requirements and steps needed to set up the database for a computer-based vibration monitoring program vary depending on the analyzer/software vendor and the system's capabilities. This section discusses the input required for such a database. However, this information should be used in conjunction with the vendor's users' manual to ensure proper implementation.

The key elements of database setup discussed in this section are analysis parameter sets, data filters (i.e., bandwidths, averaging, and weighting), limits for alerts and alarms, and data-acquisition routes.

Analysis Parameter Sets

The software used to manage the data incorporate what are referred to as analysis parameter sets (APSs). APSs define and specify machine dynamics, components, and failure modes to be monitored.

Most microprocessor-based systems permit a maximum of 256 APSs per database. This limit could be restrictive if the analyst wishes to establish a unique APS for each machine-train. To avoid this problem, APSs should be established for classes of machine-trains. For example, a group of bridle gearboxes that are identical in both design and application should share the same APS.

This approach provides two benefits. One is that it simplifies database development because one parameter set is used for multiple machine-trains. Therefore, less time is required to establish them. The other is that this approach permits direct comparison of multiple machine-trains. Since all machine-trains in a class share a common APS, the data can be directly compared. For example, the energy generated by a gear set is captured in a narrowband window established to monitor gear mesh. With the same APS, the gear mesh narrowbands can be used to compare all gear sets within that machine-train classification.

Data Filters

When selecting the bandwidth frequency range to use for data collection in a vibration-monitoring system, one might be tempted to select the broadest range available. If enough computing power were available, we could simply gather data over an infinite frequency range, analyze the data, and be assured that no impending failures were missed. However, practicalities of limited computing power prevent us from taking this approach.

Therefore, it is necessary to "filter" or screen the data we collect using our knowledge of the machinery being monitored. This is necessary to make collection, storage, and analysis of the data manageable with the equipment available. Electronic filters screen the quantity and quality of data that are collected. Mathematical filtering techniques such as resolution, averaging, and weighting are used on the data that are collected.

Bandwidth

Bandwidth frequency range settings are crucial to obtaining meaningful data for a vibration-based predictive maintenance program. Because of limits inherent with computer-based data collection and analysis systems (i.e., limited storage and data-handling capacity), these settings must be properly specified to obtain frequency data in the range generated by machine components where failures occur. Improper settings will likely yield data in frequency ranges where problems do not exist and miss critical clues to serious problems with the machinery.

Analysis Type

As discussed previously, data-collection analyzers incorporate analysis parameter sets that allow the user to control the data-gathering process. APSs provide the option of selecting either frequency analysis for fixed-speed machinery or orders analysis for variable-speed machinery.

Constant Speed: Frequency Analysis

Constant-speed machinery generates a relatively fixed set of frequency components within its signature. Therefore, specific APSs can be established to monitor using frequency analysis. Because speed is relatively constant, the location of specific frequency components (e.g., running speed) will not change greatly. Therefore, the broadband and each narrowband window can be established with a constant minimum and maximum frequency limit, which are referred to as fixed filters.

The position of these fixed filters should be set to ensure capture of information that is needed. The filter settings are determined from the speed range (i.e., no-load to full-load range) of the primary driver. In addition, the lower and upper limits of each filter should be adjusted by 10 to 15% to allow for slight variations in speed.

Variable Speed: Orders Analysis
In a variable-speed machine, the unique frequencies generated by components such as bearings and gear sets do not remain constant. As the speed changes, the unique frequency components vary in direct proportion to the speed change. For this type of machinery, the analyzer's orders analysis option is used to automatically adjust each of the filters used to set the bandwidth and narrowbands for each data set to the true machine speed.

The analyzer automatically moves the filters that designate the lower and upper limits of each narrowband window to correspond with the actual running speed at the time the data are collected. To activate this function, the technician must either manually enter the running speed or use a tachometer input to trigger data acquisition.

Boundary Conditions and Resolution
The frequency boundary conditions and resolution for the full FFT signature depend on the specific system being used. Typically, the full-signature capability of various predictive maintenance systems has a lower frequency limit of 10 Hz and an upper limit of 10 to 30 kHz. A few special low-frequency analyzers have a lower limit of 0.1 Hz, but retain the upper limit of 30 kHz. Typical resolutions are 100 to 12,800 lines.

Maximum Frequency
The dynamics of each machine-train determines the maximum frequency, F_{MAX}, that should be used for both data acquisition and analysis. The frequency must be high enough to capture and display meaningful data, but not so high that resolution is lost and meaningful data filtered out. As a general rule, when setting F_{MAX}, it is necessary to take into account the harmonics of running speed and frequencies of components such as rolling-element bearings and gear mesh. Therefore, F_{MAX} should be set to the maximum frequency encountered in any of these.

Gear Mesh
The bandwidth should include at least the second harmonic of the calculated gear-mesh frequency. This permits early detection of misalignment, gear wear, and other abnormal operating dynamics. If excessive axial thrusting is expected, the third harmonic of gear mesh also should be included. For example, helical gears are prone to generate axial movement that increases as the gears wear.

Rolling-Element Bearings
The ability to monitor rolling-element or antifriction bearing defects requires the inclusion of multiples of their rotating frequency. For example, with ball-pass inner-race bearings, the bandwidth should include the second harmonic (2×).

Running-Speed Harmonics
When setting bandwidth, at least three harmonics of running speed should be included to ensure the ability to quantify the operating-mode shape of the shaft. This is accomplished by setting F_{MAX} to at least three times the running speed.

Generally, shafts deflect from their true centerline during normal, as well as abnormal, operation. In normal operation, this deflection is slight and results in a "first mode" that creates a frequency component at the actual running speed, or first harmonic (1×). As instability increases, the shaft deflects more and more. Under abnormal operating conditions, the shaft typically deforms into either the "second mode" or "third mode." This deflection creates unique frequencies at the second (2×) and third (3×) harmonics of running speed. In addition, some failure modes (e.g., parallel misalignment) increase the energy level of one or all three of these frequencies.

Minimum Frequency
The data-collection hardware and software permit the selection of the minimum frequency, F_{MIN}, or the low-frequency cutoff below which no data are acquired and stored during the monitoring process. In most applications, however, this option should not be used.

Selecting a low-frequency cutoff does not improve resolution and is strictly an arbitrary omission of visible frequency components within a vibration signature. The FFT is calculated on a bandwidth having a lower limit of zero and an upper limit equal to the maximum frequency, F_{MAX}, which is selected by the user. The only reason for selecting a minimum frequency other than zero is to remove unneeded low-frequency components from the signature display.

Resolution
Resolution is the degree of spacing of visible frequency components in the vibration signature and is proportional to the bandwidth. The equation for resolution follows:

$$\text{Resolution} = \frac{\text{Bandwidth}}{\text{Lines of Resolution}} = \frac{F_{MAX} - F_{MIN}}{\text{Lines of Resolution}}$$

From this equation, it is apparent that the bandwidth should be as small as possible to minimize the spacing and avoid missing important data. The typical number of lines of resolution is 100 to 12,800, but it is important to make sure that the selected bandwidth and resolution include all pertinent frequency components and that they are visible in the signature.

For routine monitoring, 800 lines of resolution are recommended. Higher resolution may be needed for root-cause analysis, but this requires substantially more memory in both the analyzer and host computer. While the latter is not a major problem, higher resolution reduces the number of measurement points that can be acquired with an analyzer without transferring acquired data to the host computer. This can greatly increase the time required to complete a measurement route and should be avoided when possible.

The combination of bandwidth and lines of resolution selected for each machine-train must affect separation of the unique frequency components that represent a machine's operating dynamics. Resolution can be improved by reducing F_{MAX}, increasing the lines of resolution, or a combination of both.

For example, if the analyzer can provide a 400-line FFT, the resolution of a signature taken with a bandwidth of 0 to 20,000 Hz will be 50 Hz, or 3000 rpm, for each displayed line. The same 400-line FFT will provide a resolution of 2.5 Hz, or 150 rpm, with an F_{MAX} of 1000 Hz.

It is important to remember that the first two and last two lines of resolution are lost when the FFT is calculated. In the example just described (F_{MAX} = 1000 Hz), the first visible speed is 450 rpm and the highest visible speed is 59,700 rpm. Because the FFT always drops the first two and last two lines of resolution, the first visible frequency is three times the calculated resolution (3 × 150 = 450) and the highest visible frequency is lowered by two lines (59,700 is visible, but 59,850 and 60,000 are not shown).

Narrowbands
Analysis using narrowbands is based on specifying a series of filtered windows for a machine component or failure mode. The analyst can establish up to 12 narrowbands for each measurement point on each machine-train.

This concept reduces the manual analysis required for each data set. The analyst can scan the documentation that is generated (i.e., the exception report and trend charts) for each of the selected narrowbands to determine if further analysis is required.

Before implementing the analyzer's narrowband capability, the analyst should first understand the dynamics of each machine-train. Once this is done, establish narrowbands that bracket each of the bandwidths identifying each of the major components. At a minimum, establish a narrowband window around the following:

- Each primary running speed
- Each gear-mesh frequency (including sidebands)
- Each set of bearing frequencies
- Each blade/vane-pass frequency
- Each belt frequency

Constant-Speed Machinery
If a machine-train operates at constant speed, the best method is to set the windows using the F_{MIN} and F_{MAX} frequencies associated with the specific component. For example, a narrowband window could be established to monitor the energy generated by a gear set by defining the minimum and maximum frequencies bounding the gear mesh. The bandwidth of the narrowband should be broad enough to include the modulations, or sidebands, generated by the meshing.

For example, a gear with 50 teeth generates a gear mesh at 50 times the running speed of its shaft (50×). If the shaft turns at 100 rpm, the gear mesh frequency is 5000 cpm (also rpm). The modulations of a normal gear will occur at multiples of shaft speed (100 rpm). Therefore, in order to capture five sidebands on each side of the gear mesh frequency, the narrowband window should be established with filters set at 4500 and 5500 cpm [5000 ± (5 × 100)].

In actual practice, the narrowband filters should be somewhat greater than those in the example. Since constant-speed machines tend to have a slight variation in speed due to load variations, the narrowbands should be adjusted to compensate for these variations. In the example given previously, the limit of the lower filter should be decreased by 10% and the upper limit raised by 5% to compensate for speed variation.

Variable-Speed Machinery
Variable-speed machine-train narrowband windows should be converted to their relationship to the running speed (1x). For example, if the frequency of the ball-pass inner-race rolling-element bearing is calculated to be 5.9 times the primary shaft running speed, then the narrowband window should be set as 5.3x to 6.2x. This allows the microprocessor to track the actual bearing rotational frequency regardless of the variation in running speed.

As a general rule, the bandwidth of each narrowband should be just enough to capture the energy generated by the monitored component. Because orders analysis automatically adjusts the filters used to acquire narrowband energy data, these windows can be somewhat tighter than those used for frequency analysis.

Antialiasing Filters
Vibration data collected with a microprocessor-based analyzer can be filtered and conditioned to eliminate nonrecurring events and their associated vibration profiles. Antialiasing filters are incorporated into data-collection analyzers specifically to remove spurious signals such as impacts. While the intent behind the use of antialiasing filters is valid, their use can distort a machine's vibration profile.

Averaging

All machine-trains are subject to random, nonrecurring vibration as well as periodic vibration. Therefore, it is advisable to acquire several sets of data and average them to eliminate the spurious signals. Averaging also improves the repeatability of the data since only the continuous signals are retained.

Number of Averages
Typically, a minimum of three samples should be collected for an average. However, the factor that determines the actual number is time. One sample takes 3 to 5 sec, a four-sample average takes 12 to 20 sec, and a 1000-sample average takes 50 to 80 min to acquire. Therefore, the final determination is the amount of time that can be spent at each measurement point.

In general, three to four samples are acceptable for good statistical averaging and keeping the time required per measurement point within reason. Exceptions to this include low-speed machinery, transient-event capture, and synchronous averaging.

Table 12.1 Overlap Averaging Options

Overlap (%)	Description
0	No overlap. Data trace update rate is the same as the block processing rate. This rate is governed by the physical requirements that are internally driven by the frequency range of the requested data.
25	Terminates data acquisition when 75% of each block of new data is acquired. The last 25% of the previous sample (of the 75%) will be added to the new sample before processing is begun. Therefore, 75% of each sample is new. As a result, accuracy may be reduced by as much as 25% for each data set.
50	The last 50% of the previous block is added to a new 50% or half-block of data for each sample. When the required number of samples is acquired and processed, the analyzer averages the data set. Accuracy may be reduced to 50%.
75	Each block of data is limited to 25% new data and the last 75% of the previous block.
90	Each block contains 10% new data and the last 90% of the previous block. Accuracy of average data using 90% overlap is uncertain. Since each block used to create the average contains only 10% of actual data and 90% of a block that was extrapolated from a 10% sample, the result cannot be representative of the real vibration generated by the machine-train.

Source: Integrated Systems, Inc.

Overlap Averaging

Many of the microprocessor-based vibration monitoring systems offer the ability to increase their data-acquisition speed. This option is referred to as overlap averaging.

While this approach increases speed, it is not generally recommended for vibration analysis. Overlap averaging reduces the accuracy of the data and must be used with caution. Its use should be avoided except where fast transients or other unique machine-train characteristics require an artificial means of reducing the data-acquisition and processing time.

When sampling time is limited, a better approach is to reduce or eliminate averaging altogether in favor of acquiring a single data block, or sample. This reduces the acquisition time to its absolute minimum. In most cases, the single-sample time interval is less than the minimum time required to obtain two or more data blocks using the maximum overlap-averaging sampling technique. In addition, single-sample data are more accurate.

Table 12.1 describes overlap-averaging options. Note that the approach described in this table assumes that the vibration profile of monitored machines is constant.

Window Selection: Weighting or Signal Conditioning

The user can select the type of signal conditioning, or weighting used to display the vibration signature. For routine monitoring, the Hanning window should be selected. The flat-top window should be used for waterfall analysis, which modifies the profile of each frequency component so that the true amplitude is displayed.

Hanning Correction

The Hanning correction provides the best capture of the individual frequency components of a signature. However, this weighting factor may distort the actual amplitude of the frequency components. Nevertheless, it is used for routine monitoring using FFT analysis.

Flat-Top Weighting

Flat-top weighting provides the best representation of the actual amplitude of each frequency component. However, it may distort the actual location (i.e., frequency) of each component.

Flat-top weighting is useful when doing waterfall analysis. Even though the actual location of each frequency component may be slightly out of position, the profile is more visible when closely packed into a waterfall display. However, it is not normally used for single-channel FFT analysis.

Alert/Alarm Limits

All microprocessor-based predictive maintenance systems are designed to automatically evaluate degradation of the machine-train by monitoring the change in amplitude of vibration using trending techniques. By establishing a series of alert/alarm limits in the database, the system can automatically notify the analyst that degradation is occurring. At least three levels of alert/alarms should be established: 1) low-limit alert, (2) maximum rate of change, and (3) absolute fault.

Low-Limit Alert

The first alert, the low-limit alert, should be set at the lowest vibration amplitude that will be encountered from a normally operating machine-train. This value is needed to ensure that valid data are taken with the microprocessor. If this minimum amplitude is not reached, the system alerts the operator, who can retake or verify the data point. Low-limit selection is arbitrary, but should be set slightly above the noise floor of the specific microprocessor used to acquire data.

Maximum Rate of Change Alert

The second alert, the maximum rate of change alert, is used to automatically notify the operator that, based on statistical data, the rate of degradation has increased above the preselected norm. Since the vibration amplitudes of all machine-trains increase as normal wear occurs, the statistical rate of this normal increase should be trended. A drastic change in this rate is a major indication that a problem is developing.

The system should be able to establish the norm based on trends developed over time. However, the analyst must establish the level of deviation that triggers the alarm. The level of deviation in rate depends on the mechanical condition of each machine-train. For a new machine in good operating condition, the limit is typically set at two times the norm. However, this must be adjusted based on the actual baseline of the machine. Note that it is better to set the limit too low initially and adjust it later.

Absolute-Fault Alarm

The third limit, the absolute-fault alarm, is the most critical of the alert/alarm limits. When this limit is reached, the probability of catastrophic failure within 1000 operating hours is greater than 90%.

Absolute-fault limits are typically based on industrial standards for specific classifications of machinery. Generally, these standards are based on a filtered broadband limit and are not adjusted for variables such as speed, load, or mounting configuration. However, vibration amplitude and its severity depend on speed and load. Therefore, alert/alarm limits must be adjusted for variations in both of these critical factors.

Effect of Speed on Limit
Vibration-severity charts, such as the Rathbone chart illustrated in Figure 12.1, provide a basis for establishing the absolute-fault limit for machinery. Note, however, that the Rathbone chart does not adjust the maximum limit for speed—something that can cause a serious problem in most industrial applications.

A 2-mil shaft displacement at 600 rpm is acceptable for most applications, but the same displacement at 1800 rpm is considered severe. Therefore, alert/alarm limits must be established based on the actual speed range of each machine-train. When casing severity is used (i.e., data are taken from the bearing caps rather than actual shaft displacement), the limits can be grouped into three basic speed ranges: less than 299 rpm, 300 to 1199 rpm, and 1200 through 3600 rpm. Tables 12.2–12.4 provide the alert/alarm limits for each speed range.

Effect of Load on Limit
Load has a direct impact on the vibration energy generated by a machine-train. For example, a centrifugal compressor operating at full load will have a lower level of vibration than the same compressor operating at 50% load. This change in vibration energy is the direct result of a corresponding change in the spring stiffness of the rotating element under varying load conditions.

The rated, or design, load of a machine establishes the following elements: (1) spring constant, (2) stiffness of the rotating element, and (3) damping coefficient of its sup-

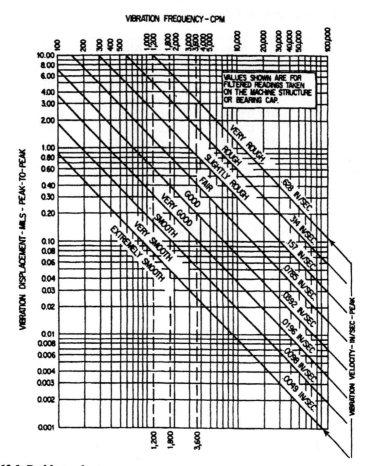

Figure 12.1 Rathbone chart.

Table 12.2 Alarm Limits for 1200 rpm and Higher

Bandwidth	Alert	Alarm	Absolute Fault
Overall	0.15 IPS-PK	0.30 IPS-PK	0.628 IPS-PK
1× Narrowband	0.10 IPS-PK	0.20 IPS-PK	0.40 IPS-PK
2× Narrowband	0.05 IPS-PK	0.10 IPS-PK	0.20 IPS-PK
3× Narrowband	0.04 IPS-PK	0.08 IPS-PK	0.15 IPS-PK
1× Gear mesh	0.05 IPS-PK	0.10 IPS-PK	0.2 IPS-PK
Rolling-element bearing	0.05 IPS-PK	0.10 IPS-PK	0.2 IPS-PK
Blade/vane pass	0.05 IPS-PK	0.10 IPS-PK	0.2 IPS-PK

Source: Integrated Systems, Inc.

Table 12.3 Alarm Limits for 300 to 1199 rpm

Bandwidth	Alert	Alarm	Absolute Fault
Overall	0.10 IPS-PK	0.15 IPS-PK	0.30 IPS-PK
1× Narrowband	0.05 IPS-PK	0.10 IPS-PK	0.20 IPS-PK
2× Narrowband	0.02 IPS-PK	0.05 IPS-PK	0.10 IPS-PK
3× Narrowband	0.01 IPS-PK	0.03 IPS-PK	0.06 IPS-PK
1× Gear mesh	0.02 IPS-PK	0.04 IPS-PK	0.08 IPS-PK
Rolling-element bearing	0.03 IPS-PK	0.05 IPS-PK	0.10 IPS-PK
Blade/vane pass	0.03 IPS-PK	0.05 IPS-PK	0.10 IPS-PK

Source: Integrated Systems, Inc.

Table 12.4 Alarm Limits for 299 rpm and Below

Bandwidth	Alert	Alarm	Absolute Fault
Overall	0.05 IPS-PK	0.10 IPS-PK	0.20 IPS-PK
1× Narrowband	0.03 IPS-PK	0.06 IPS-PK	0.12 IPS-PK
2× Narrowband	0.01 IPS-PK	0.03 IPS-PK	0.06 IPS-PK
3× Narrowband	0.01 IPS-PK	0.02 IPS-PK	0.04 IPS-PK
1× Gear mesh	0.02 IPS-PK	0.03 IPS-PK	0.05 IPS-PK
Rolling-element bearings	0.02 IPS-PK	0.03 IPS-PK	0.05 IPS-PK
Blade/vane pass	0.02 IPS-PK	0.03 IPS-PK	0.05 IPS-PK

Source: Integrated Systems, Inc.

port system. Therefore, when load varies from design, the stiffness of the rotor and the rotor-support system also must change.

This change in vibration energy can be clearly observed in trend data acquired from machine-trains. A sawtooth trend is common to most predictive maintenance programs, which can be directly attributed to variations in load. The only way to compensate for load variations is to track the actual load associated with each data set.

Mounting Configuration and Operating Envelope
Industrial standards, such as the Rathbone severity chart, assume that the machine is rigidly mounted on a suitable concrete foundation. Machines mounted on deck-plate or on flexible foundations have higher normal vibration profiles and cannot be evaluated using these standards.

In addition, industrial standards assume a normal operating envelope. All machines and process systems have a finite operating range and must be operated accordingly. Deviations from either the design operating envelope or from best operating practices will adversely affect the operating condition and vibration level of the machine.

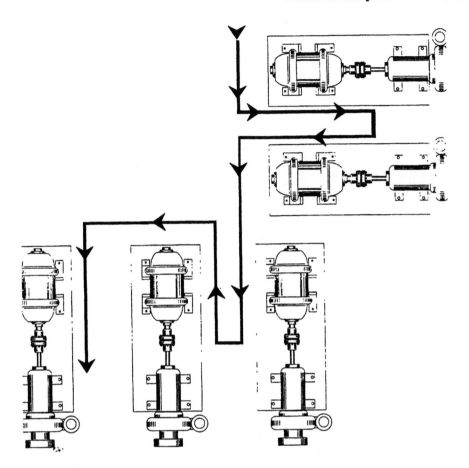

Figure 12.2 Typical measurement route.

Data-Acquisition Routes

Most computer-based systems require data-acquisition routes to be established as part of the database setup. These routes specifically define the sequence of measurement points and, typically, a route is developed for each area or section of the plant. With the exception of limitations imposed by some of the vibration monitoring systems, these routes should define a logical walking route within a specific plant area. A typical measurement is shown in Figure 12.2.

Most of the computer-based systems permit rearrangement of the data-acquisition sequence in the field. They provide the ability to skip through the route until the appropriate machine, or the measurement point of a machine, is located. However, this manual adjustment of the preprogrammed route is time consuming and should be avoided whenever possible.

Chapter 13

VIBRATION DATA ACQUISITION

Limitations on data acquisition arise due to the use of portable handheld, micropro-cessor-based analyzers to obtain data. Limits are cost, weight restrictions, and the fact that they are generally designed for a technician to manually take a series of single measurements directly from individual machine-trains or machine-train components. Therefore, this discussion is limited to the best practices for acquiring single-channel, frequency-domain data using portable, handheld analyzers. It does not address multi-channel or other nonserial data-acquisition techniques.

Data-acquisition and vibration-detection equipment (i.e., analyzers and transducers) are critical factors that determine the success or failure of a vibration monitoring and analysis program. Their accuracy, proper usage, and mounting determine if valid data are collected. An optimum program is based on the accuracy and repeatability of the data, both of which are negatively affected by the use of the wrong transducer or mounting technique.

TRANSDUCERS

The three basic types of transducers that can be used for monitoring the mechanical condition of plant machinery are (1) displacement probes (measures movement), (2) velocity transducers (measures energy due to velocity), and (3) accelerometers (mea-sures force due to acceleration). Each has specific applications in a monitoring pro-gram, while each also has limitations.

MEASUREMENT ORIENTATION

Most vibration-monitoring programs rely on data acquired from machine housings or bearing caps. The only exceptions are applications that require direct measurement of

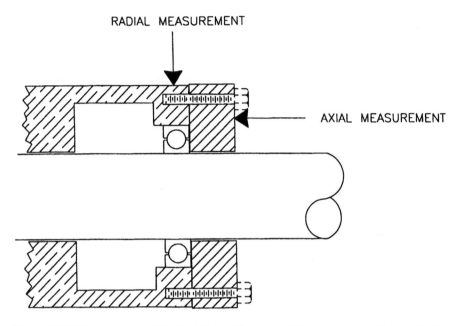

Figure 13.1 Measurement points should provide shortest direct mechanical link to shaft.

actual shaft displacement to gain an accurate picture of the machine's dynamics. Transducers used to acquire the data are mounted either radially or axially.

MEASUREMENT LOCATIONS

Each measurement point, typically located on the bearing housing or machine casing, should provide the shortest direct mechanical link to the shaft. Figure 13.1 illustrates such a location oriented in both the axial and radial planes. If a transducer is not mounted in an appropriate location, the data will be distorted by noise such as fluid flow in the bearing reservoir or through the machine.

Measurement locations should be permanently marked to ensure repeatability of data. If transducers are permanently mounted, the location can be marked with a center-punch, paint, or any other method that identifies the point. The following sections give the recommended locations and orientations of measurement points for the following common machines or machine components: compressors, electric motors, fans and blowers, gearboxes, process rolls, and pumps.

Compressors

In most cases, measurement point locations for compressors are identical to those of pumps and fans. If a compressor is V-belt driven, the primary measurement point should be in a plane opposing the side load created by the belts on the inboard and outboard bearings. The secondary point should be at 90 degrees to the primary point.

At least one axial measurement point should be located on each compressor shaft. Axial data are helpful in identifying and quantifying thrust (i.e., induced) loads created by both the process and any potential compressor-element problems, such as imbalance, cracked blade, etc.

In applications where numerous compressors are in proximity, an additional measurement point on the base is useful for identifying structural resonance or cross-talk between the units.

Centrifugal

The two major types of centrifugal compressors used in industrial applications are in-line and bullgear compressors.

In-Line Centrifugal Compressors
Measurement locations for in-line centrifugal compressors should be based on the same logic as discussed for pumps. Impeller design and orientation, as well as the inlet and discharge configurations, are the dominant reasons for point location. (Figure 11.5 illustrates a typical multistage, in-line compressor.)

The in-line impeller configuration generates high axial thrusting, which increases the importance of the axial (Z-axis) measurement point. That point should be on the fixed bearing and oriented toward the driver.

In addition, this type of compressor tends to have both the suction and discharge ports on the same side of the compressor's housing. As a result, there is a potential for aerodynamic instability within the compressor. Orientation of the primary (X-axis) radial measurement point should be opposite the discharge port and oriented toward the discharge. The secondary (Y-axis) radial point should be in the direction of shaft rotation and 90 degrees from the primary radial point.

Bullgear Compressors
Because of the large number of these machines being manufactured, proper locations for displacement, or proximity, probes have been established by the various machine manufacturers. Nearly all of these compressors are supplied by the original equipment manufacturer (OEM) with one or two proximity probes already mounted on each pinion shaft and, sometimes, one probe on the bullgear shaft. These probes can be used to obtain vibration data with the microprocessor-based portable analyzers. However, they must be augmented with casing measurements acquired from suitable accelerometers. This is necessary because there are two problems with the proximity data.

First, most of the OEM-supplied data-acquisition systems perform signal conditioning on the raw data acquired from the probes. If the conditioned signal is used, there is a bias in the recorded amplitude. This bias may increase the raw-data signal by 30 to 50%. If data from the proximity probes are to be used, it is better to acquire it before signal conditioning. This can be accomplished by tapping into the wiring between the probe and display panel.

The second problem is data accuracy. The pinions on most bullgear compressors rotate at speeds between 20,000 and 75,000 rpm. While these speeds, for the most part, are within the useful range of a proximity probe (600 to 60,000 rpm), the frequencies generated by common components (i.e., tilting-pad bearings and impeller vane-pass) are well outside this range. In addition, proximity probes depend on a good sight picture, which means a polished shaft that has no endplay or axial movement. Neither of these conditions is present in a bullgear compressor.

Primary (X-axis) and secondary (Y-axis) radial measurements should be acquired from both bearings on the bullgear shaft. If the shaft has Babbitt bearings, it is a good practice to periodically acquire four radial readings, one at each quadrant of the bearing, to determine the load zones of the bearing. Normal vertical and horizontal locations are acceptable for the routine readings, but primary (X-axis) measurement points should be in the horizontal plane (i.e., 90 degrees from vertical in the direction of rotation). For clockwise rotation, the primary should be on the right side and for counterclockwise on the left.

Because a bullgear compressor incorporates a large helical gear, the shaft displays moderate to high axial thrusting. Therefore, an axial (Z-axis) measurement point should be acquired from the thrust (outboard) bearing oriented toward the driver.

The pinion shafts in this type of compressor are inside the housing. As a result, it is difficult to obtain radial measurements directly. A cross-sectional drawing of the compressor is required to determine the best location and orientation for the measurement points.

Positive Displacement

Two major types of positive-displacement compressors are used in industrial applications: reciprocating and screw.

Reciprocating Compressors
Limitations of the frequency-domain analysis prevent total analysis of reciprocating compressors. It is limited to the evaluation of the rotary forces generated by the main crankshaft. Therefore, time-domain and phase analysis are required for complete diagnostics.

The primary (X-axis) radial measurement point should be located in a plane opposite the piston and cylinder. Its orientation should be toward the piston's stroke. This orientation provides the best reading of the impacts and vibration profile generated by the reversing linear motion of the pistons. The secondary (Y-axis) radial measurement

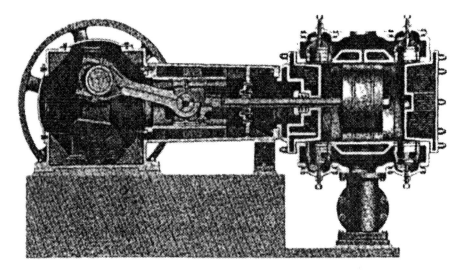

Figure 13.2 Typical cross-section of a reciprocating compressor.

point should be spaced at 90 degrees to the primary point and in the direction of rotation at the main crankshaft. This configuration should be used for all accessible main crankshaft bearings. Figure 13.2 provides a typical cross-section of a reciprocating compressor, which will assist in locating the best measurement points. Similar drawings are available for most compressors and can be obtained from the vendor. There should be little axial thrusting of the main crankshaft, but an axial (Z-axis) measurement point should be established on the fixed bearing, oriented toward the driver.

If the vibration analyzer permits acquisition of time-domain data, additional time-waveform data should be obtained from the intermediate guide as well as the inlet and discharge valves. The intermediate guide is located where the main crankshaft lever arm connects to the piston rod. Time waveforms from these locations detect any binding or timing problems that may exist in the compressor.

Screw Compressors
Figure 11.10 illustrates a typical single-stage screw compressor. Radial measurements should be acquired from all bearing locations in the compressor. The primary bearing locations are the inboard, or float, bearings on the driver side of the compressor housing and the fixed bearing located on the outboard end of each shaft.

In most cases, the outboard bearings are not directly accessible and measurement points must be located on the compressor's casing. Extreme care must be taken to ensure proper positioning. A cross-sectional drawing facilitates selection of the best, most direct mechanical link to the these bearings.

The primary (X-axis) radial measurement point should be located opposite the mesh of the rotors and oriented toward the mesh. In the illustration, the primary point is on the top of the housing and oriented in the downward direction. The secondary (Y-axis) radial measurement point should be in the direction of rotation and 90 degrees from the primary.

Because of the tendency for screw compressors to generate high axial vibration when subjected to changes in process conditions, the axial (Z-axis) measurement point is essential. The ideal location for this point is on the outboard, or fixed, bearing and oriented toward the driver. Unfortunately, this is not always possible. The outboard bearings are fully enclosed within the compressor's housing and an axial measurement cannot be obtained at these points. Therefore, the axial measurement must be acquired from the float, or inboard, bearings. While this position captures the axial movement of the shaft, the recorded levels are lower than those acquired from the fixed bearings.

Electric Motors

Both radial (X- and Y-axis) measurements should be taken at the inboard and outboard bearing housings. Orientation of the measurements is determined by the anticipated induced load created by the driven units. The primary (X-axis) radial measurement should be positioned in the same plane as the worst anticipated shaft displacement. The secondary (Y-axis) radial should be positioned at 90 degrees in the direction of rotation to the primary point and oriented to permit vector analysis of actual shaft displacement.

Horizontal motors rely on a magnetic center generated by its electrical field to position the rotor in the axial (Z-axis) plane between the inboard and outboard bearings. Therefore, most electric motors are designed with two float bearings instead of the normal configuration incorporating one float and one fixed bearing. Vertical motors should have an axial (Z-axis) measurement point at the inboard bearing nearest the coupling and oriented in an upward direction. This data point monitors the downward axial force created by gravity or an abnormal load.

Electric motors are not designed to absorb side loads, such as those induced by V-belt drives. In applications where V-belts or other radial loads are placed on the motor, the primary radial transducer (X-axis) should be oriented opposite the direction of induced load and the secondary radial (Y-axis) point should be positioned at 90 degrees in the direction of rotation. If, for safety reasons, the primary transducer cannot be positioned opposite the induced load, the two radial transducers should be placed at 45 degrees on either side of the load plane created by the side load.

Totally enclosed, fan-cooled, and explosion-proof motors present some difficulty when attempting to acquire data on the outboard bearing. By design, the outboard bearing housing is not accessible. The optimum method of acquiring data is to permanently mount a sensor on the outboard-bearing housing and run the wires to a convenient data-acquisition location. If this is not possible, the X-Y data points should be as close as possible to the bearing housing. Ensure that there is a direct mechanical path to the outboard bearing. The use of this approach results in some loss of signal strength from motor-mass damping. Do not obtain data from the fan housing.

Fans and Blowers

If a fan is V-belt driven, the primary measurement point should be in a plane opposing the side load created by the belts on the inboard and outboard bearings. The secondary point should be at 90 degrees to the primary in the direction of rotation.

Bowed shafts caused by thermal and mechanical effects create severe problems on large fans, especially overhung designs. Therefore, it is advantageous to acquire data from all four quadrants of the outboard bearing housing on overhung fans to detect this problem.

At least one axial measurement point should be located on each fan shaft. This is especially important on fans that are V-belt driven. Axial data are helpful in identifying and quantifying thrust (induced) loads created by the process and any potential fan element problems such as imbalance, cracked blade, etc.

In applications where numerous fans are in proximity, an additional measurement point on the base is useful for identifying structural resonance or cross-talk between the fans.

Gearboxes

Gearbox measurement point orientation and location should be configured to allow monitoring of the normal forces generated by the gear set. In most cases, the separating force, which tends to pull the gears apart, determines the primary radial measurement point location. For example, a helical gear set generates a separating force that is tangential to a centerline drawn through the pinion and bullgear shafts. The primary (X-axis) radial measurement point should be oriented to monitor this force and a secondary (Y-axis) radial should be located at 90 degrees to the primary. The best location for the secondary (Y-axis) radial is opposite the direction of rotation. In other words, the secondary leads the primary transducers.

With the exception of helical gears, most gear sets should not generate axial or thrust loads in normal operation. However, at least one axial (Z-axis) measurement point should be placed on each of the gear shafts. The axial point should be located at the fixed, or thrust, bearing cap and oriented toward the gearbox.

In complex gearboxes, it may be difficult to obtain radial measurements from the intermediate or idler shafts. In most cases, these intermediate shafts and their bearings are well inside the gearbox. As a result, direct access to the bearings is not possible. In these cases, the only option is to acquire axial (Z-axis) readings through the gearbox housing. A review of the cross-sectional drawings allows the best location for these

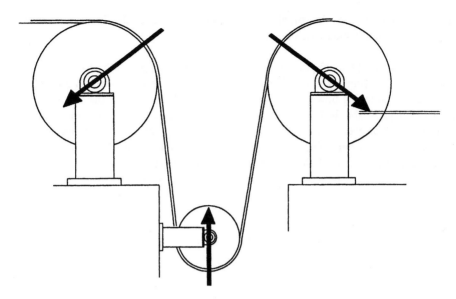

Figure 13.3 Typical process-roll configuration and wrap-force vectors.

measurements to be determined. The key is to place the transducer at a point that will provide the shortest, direct link to the intermediate shaft.

Process Rolls

Process rolls are widely used by industry. As with other machine components, two radial (X- and Y-axis) and one axial (Z-axis) measurement should be acquired from each roll. However, the orientation of these measurement points is even more critical for process rolls than for some of the other machine components.

The loading on each roll is generated by the belt, wire mesh, and/or transported product. The amount and distribution of the load varies depending on the wrap of the carried load. *Wrap* refers to the angular distance around the roll that touches the belt, wire mesh, or product. In most conveyor systems, the load is relatively uniform and is in a downward direction. In this case, the traditional vertical, horizontal, and axial mounting positions are acceptable.

Figure 13.3 represents a typical process-roll configuration. The arrows indicate the force vectors generated by the wire, belt, or product wrap around these rolls. The left roll has a force vector at 45 degrees down to the left; the right roll has a mirror image force vector; and the bottom roll has a vertical vector.

The primary (X-axis) radial measurement for the bottom roll should be in the vertical plane with the transducer mounted on top of the bearing cap. The secondary radial (Y-axis) measurement should be in the horizontal plane facing upstream of the belt.

Since the belt carried by the roll also imparts a force vector in the direction of travel, this secondary point should be opposite the direction of belt travel.

The ideal primary (X-axis) point for the top right roll is opposite the force vector. In this instance, the primary radial measurement point should be located on the right of the bearing cap facing upward at a 45-degree angle. Theoretically, the secondary (Y-axis) radial point should be at 90 degrees to the primary on the bottom-left of the bearing cap. However, it may be difficult, if not impossible, to locate and access a measurement point here. Therefore, the next best location is at 45 degrees from the anticipated force vector on the left of the bearing cap. This placement still provides the means to calculate the actual force vector generated by the product.

Pumps

Appropriate measurement points vary by type of pump. In general, pumps can be classified as centrifugal or positive displacement, and each of these can be divided into groups.

Centrifugal Pumps

The location of measurement points for centrifugal pumps depends on whether the pump is classified as end suction or horizontal splitcase.

End Suction Pumps

Figure 13.4 illustrates a typical single-stage, end-suction centrifugal pump. The suction inlet is on the axial centerline, while the discharge may be either horizontal or vertical. In the illustration, the actual discharge is horizontal and is flanged in the vertical downstream.

The actual discharge orientation determines the primary radial (X-axis) measurement point. This point must be oriented in the same plane as the discharge and opposite the direction of flow. In the illustration, the primary point should be in the horizontal plane facing the discharge.

Restrictions or other causes of back-pressure in the discharge piping deflect the shaft in the opposite direction. Referring back to the illustration, the shaft would be deflected toward the front of the picture. If the discharge were vertical and in the downward direction, the primary radial measurement point would be at the top of the pump's bearing cap looking downward.

A second radial (Y-axis) measurement point should be positioned at 90 degrees to the primary in a plane that captures secondary shaft deflection. For the pump illustrated in Figure 13.4, the secondary (Y-axis) radial measurement point is located on top of the pump's bearing cap and oriented downward. Since the pump has a clockwise rotation, back-pressure in the discharge piping forces the shaft both downward and horizontally toward the front of the picture.

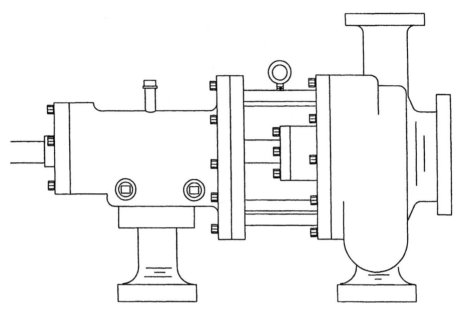

Figure 13.4 Typical end-suction, single-stage centrifugal pump.

Because this type of pump is susceptible to axial thrusting, an axial (Z-axis) measurement point is essential. This point should be on the fixed bearing housing oriented toward the driver.

Horizontal Splitcase Pumps

The flow pattern through a horizontal splitcase pump is radically different than that through an end-suction pump. Inlet and discharge flow are in the same plane and almost directly opposite one another. This configuration, illustrated in Figure 13.5, greatly improves the hydraulic-flow characteristics within the pump and improves its ability to resist flow-induced instability.

The location of the primary (X-axis) radial measurement point for this type of pump is in the horizontal plane and on the opposite side from the discharge. The secondary (Y-axis) radial measurement point should be 90 degrees to the primary point and in the direction of rotation. If the illustrated pump has a clockwise rotation, the measurement point should be on top, oriented downward. For a counterclockwise rotation, it should be on bottom, oriented upward.

Single-stage pumps generate some axial thrusting due to imbalance between the discharge and inlet pressures. The impeller design provides a means of balancing these forces, but it cannot absolutely compensate for the difference in the pressures. As a result, there will be some axial rotor movement. In double volute, or multistage,

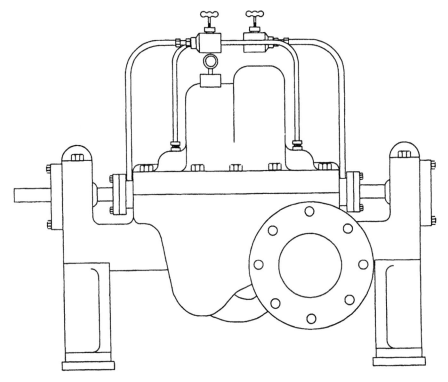

Figure 13.5 Typical horizontal splitcase pump.

pumps, two impellers are positioned back to back. This configuration eliminates most of the axial thrusting when the pump is operating normally.

An axial (Z-axis) measurement point should be located on the fixed bearing housing. It should be oriented toward the driver to capture any instability that may exist.

Multistage Pumps

Multistage pumps may be either end-suction or horizontal splitcase pumps. They have two basic impeller configurations, in-line or opposed, as shown in Figure 13.6. In-line impellers generate high thrust loads.

The impeller configuration does not alter the radial measurement locations discussed in the preceding sections. However, it increases the importance of the axial (Z-axis) measurement point. The in-line configuration drastically increases the axial loading on the rotating element and, therefore, the axial (Z-axis) measurement point is critical. Obviously, this point must be in a location that detects axial movement of the shaft. However, since large, heavy-duty fixed bearings are used to withstand the high thrust loading generated by this design, direct measurement is difficult. A cross-sectional

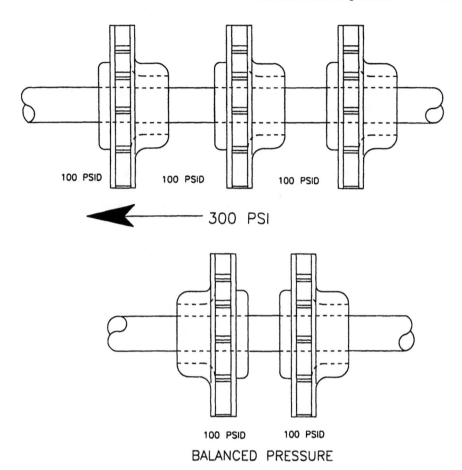

Figure 13.6 In-line and opposed impellers on multistage pumps.

drawing of the pump may be required to locate a suitable location for this measurement point.

Positive-Displacement Pumps

Positive-displacement pumps can be divided into two major types: rotary and reciprocating. All rotary pumps use some form of rotating element, such as gears, vanes, or lobes to increase the discharge pressure. Reciprocating pumps use pistons or wobble plates to increase the pressure.

Rotary Pumps

Locations of measurement points for rotary positive-displacement pumps should be based on the same logic as in-line centrifugal pumps. The primary (X-axis) radial measurement should be taken in the plane opposite the discharge port. The secondary

(*Y*-axis) radial should be at 90 degrees to the primary and in the direction of the rotor's rotation.

Since most rotary positive-displacement pumps have inlet and outlet ports in the same plane and opposed, there should be relatively little axial thrusting. However, an axial (Z-axis) measurement should be acquired from the fixed bearing, oriented toward the driver.

Reciprocating Pumps

Reciprocating pumps are more difficult to monitor because of the combined rotational and linear motions that are required to increase the discharge pressure. Measurement point location and orientation should be based on the same logic as that of reciprocating compressors.

Chapter 14

TRENDING ANALYSIS

Long-term vibration trends are a useful diagnostic tool. Trending techniques involve graphically comparing the total energy, which is the sum of the frequency components' amplitude over some consistent, user-selected frequency range (i.e., F_{MIN} to F_{MAX}), over a long period of time to get a historical perspective of the vibration pattern. Plots of this sum against time (e.g., days) provide a means of quantifying the relative condition of the monitored machine (see Figure 14.1). Most predictive maintenance systems provide automatic-trending capabilities for recorded data. This is not to be confused with time-domain plots, which are instantaneous measures of total vibration amplitude plotted against time measured in seconds.

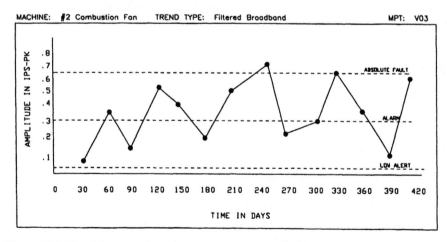

Figure 14.1 Trend data are plotted versus time and provide historical trends.

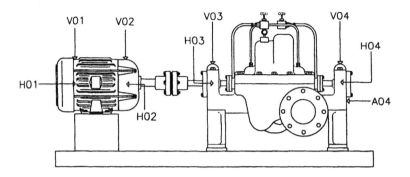

POINT	VALUE (IPS-Peak)	POINT	VALUE (IPS-Peak)
V01	0.12	H01	0.11
V02	0.15	H02	0.12
V03	0.15	H03	0.10
V04	0.12	H04	0.09
A04	0.14		

Figure 14.2 Typical broadband measurements.

Used properly, this feature greatly enhances a predictive maintenance program. The real value of trending techniques is that they provide the capability of automatically scanning large amounts of data (both broadband and narrowband) and reporting any change in preselected values.

TYPES OF TRENDS

The three primary categories of trends are broadband, narrowband, and combinations of the two.

Broadband Trends

Most microprocessor-based vibration-monitoring systems acquire and record a filtered broadband energy level for each data point included in the program. The bandwidth of the energy band is determined by the minimum, F_{MIN}, and maximum, F_{MAX}, frequencies that were established as part of the database setup. In most applications, the minimum frequency should be zero, but the maximum varies, depending on the specific machine-train. Figure 14.2 illustrates typical broadband data.

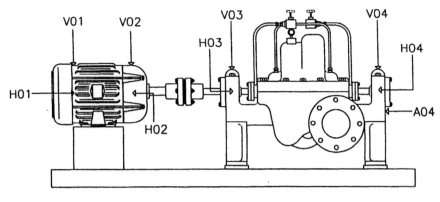

POINT	SHAFT INSTABILITY	VANEPASS	BEARING DEFECTS
V01	0.10	0.09	0.001
V02	0.10	0.09	0.002
V03	0.13	0.10	0.002
V04	0.11	0.09	0.010

Figure 14.3 Narrowband data.

Broadband data cannot be used to identify specific machine components (e.g., bearing, gears) or failure modes (e.g., imbalance, misalignment). The data acquired using broadband filters are limited to the total energy value contained within the user-selected frequency window or bandwidth, F_{MIN} to F_{MAX}.

At best, broadband energy provides a gross approximation of the machine's condition and its relative rate of degradation. Since the only available data are overall energy values, broadband data do not provide enough detail to permit diagnosis of machine-train condition. Without discrete identification of the specific frequencies that make up the overall energy, the failure mode or failing component cannot be determined.

Narrowband Trends

Like broadband data, narrowband data also reflect the total energy, but it reflects a more restricted user-selected range or window. Narrowband monitoring generally is used to trend and evaluate one selected machine-train component rather than several. Filtered narrowband windows are typically set up around the unique frequency components generated by specific machine-train components so that the energy in each filtered window can be directly attributed to that specific machine component. However, even though narrowband analysis improves the diagnostic capabilities of a predictive maintenance program, it is not possible to isolate and identify specific failure modes within a machine-train. Figure 14.3 illustrates the added information provided by narrowband data.

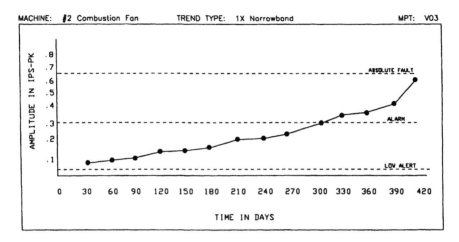

Figure 14.4 Narrowband trends provide energy histories of specific components.

Figure 14.4 illustrates narrowband data trends. In addition to the overall or broadband energy, narrowband trends indicate the relative energy in select machine-train components. In effect, this type of analysis is a series of mini-overall energy readings.

Composite Trends

Most microprocessor-based systems permit composite trending (i.e., simultaneous displays) of both filtered broadband and narrowband data. Figure 14.5 illustrates a composite trend that includes both broadband and narrowband data. This type of plot is quite beneficial because it permits the analyst to track the key indicators of machine condition on one plot.

EVALUATION METHODS

Trend data can be used in the following ways: (1) to compare with specific reference values, (2) mode-shape comparisons, and (3) cross-machine comparisons.

Comparison to Reference Values

Three types of reference-value comparisons are used to evaluate trend data: baseline data, rate of change, and industrial standards.

Baseline Data

A series of baseline or reference data sets should be taken for each machine-train included in a predictive maintenance program. These data sets are necessary for future use as a reference point for trends, time traces, and FFT signatures that are col-

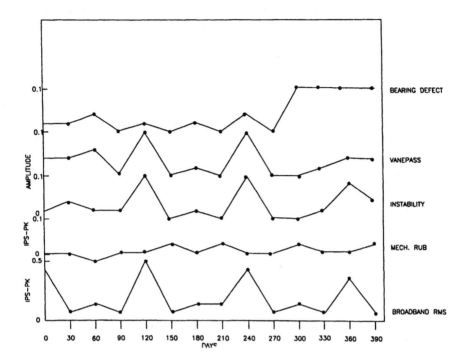

Figure 14.5 Combined trends (i.e., composite) provide both broadband and narrowband data.

lected over time. Such baseline data sets must be representative of the normal operating condition of each machine-train to have value as a reference.

Three criteria are critical to the proper use of baseline comparisons: reset after maintenance, proper identification, and process envelope.

Reset After Maintenance
The baseline data set must be updated each time the machine is repaired, rebuilt, or when any major maintenance is performed. Even when best practices are used, machinery cannot be restored to as-new condition when major maintenance is performed. Therefore, a new baseline or reference data set must be obtained following these events.

Proper Identification
Each reference or baseline data set must be clearly and completely identified. Most vibration-monitoring systems permit the addition of a label or unique identifier to any user-selected data set. This capability should be used to clearly identify each baseline data set.

In addition, the data set label should include all defining information. For example, any rework or repairs made to the machine should be identified. If a new baseline data set is obtained after the replacement of a rotating element, this information should be included in the descriptive label.

Process Envelope
Since variations in process variables, such as load, have a direct effect on the vibration energy and signature generated by a machine-train, the actual operating envelope for each baseline data set must also be clearly identified. If this step is omitted, direct comparison of other data to the baseline will be meaningless. The label feature in most vibration monitoring systems permits tagging the baseline data set with this additional information.

Rate of Change

Rate of change is the most often used trend analysis. Because most of the microprocessor-based systems provide the ability to automatically display both broadband and narrowband data trends, analysts tend to rely on this means of comparative analysis.

Rate of change is a valid means of defining the relative condition of rotating machinery. As a general rule, there must be a change in mechanical condition before there can be a change in the vibration energy generated by a machine. Therefore, monitoring the rate that the energy levels change, either up or down, is a useful tool (see Figure 14.6). *Caution:* Broadband and narrowband data must be normalized for changes in load before being valid. Normal variations in machine load destroy the validity of non-normalized trend data and little can be gained from its use.

Industrial Standards

There are a number of published standards that define acceptable levels of vibration in machinery. These standards are valuable reference tools, but they must be clearly understood and properly used. Industrial-standard data can be obtained from a Rathbone chart and from the American Petroleum Institute.

Rathbone Vibration-Severity Chart
The Rathbone chart provides levels of vibration severity that range from extremely smooth, which is the best possible operating condition, to very rough or absolute-fault limit, which is the maximum level where a machine can operate (see Figure 14.7).

This chart is useful, although it is often misused. Four factors must be understood before using the chart:

1. Data included in the Rathbone chart are valid for machines with running speeds between 1200 and 3600 rpm. The chart cannot be used for low-speed or turbomachinery.

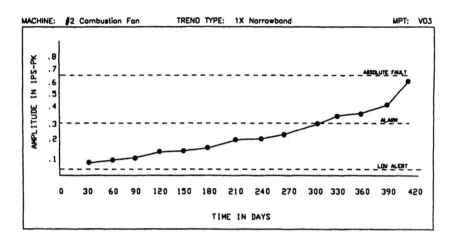

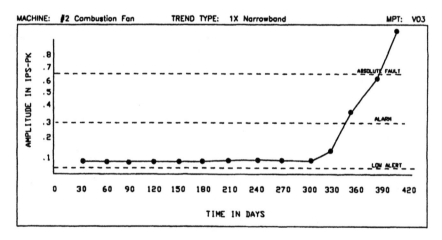

Figure 14.6 Rate of change of trend data indicates condition change.

2. The data presented in the chart are relative vibration levels (i.e., taken from a bearing pedestal using either an accelerometer or velocity probe) in inches per second (in./sec) peak.

3. Data are peak values of velocity (in./sec) for a filtered broadband from 10 to 10,000 Hz.

4. The severity levels are relative, not absolute. For example, when a machine reaches the absolute-fault limit, it has a 90% probability of failure within its next 1000 hours of operation (i.e., it is not going to fail tomorrow).

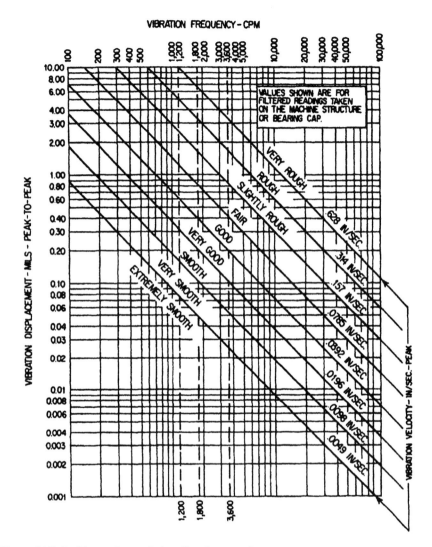

Figure 14.7 Rathbone chart-relative vibration severity.

American Petroleum Institute Standards

The American Petroleum Institute (API) has established standards for vibration levels. Unlike the Rathbone chart, which presents relative vibration data, the API standards are actual shaft displacement as measured with a displacement probe. Based on the API data, acceptable vibration can be defined by:

$$\text{Vibration Severity} = \sqrt{\frac{12,000}{\text{rpm}}} \text{ mils, peak-to-peak}$$

The API equivalent of the absolute-fault limit can be defined by:

$$\text{Absolute-fault Limit} = 1.3 \times \sqrt{\frac{12,000}{\text{rpm}}} \text{ mils, peak-to-peak}$$

The API standards are reasonable for turbomachinery, but are unacceptable for lower speed machines. The standards are applicable for rotating machinery with speeds above 1800 rpm; marginal for speeds between 600 and 1800 rpm; and should not be used for speeds below 600 rpm.

Mode Shape (Shaft Deflection)

A clear understanding of the mode shape, or shaft deflection, of a machine's rotating element is a valuable diagnostic tool. Both broadband and narrowband filtered energy windows can be used at each measurement point and orientation across the machine. The resultant plots, one in the vertical plane and one in the horizontal plane, provide an approximation of the mode shape of the complete machine and its rotating element.

Unfortunately, these plots must be developed manually. The microprocessor-based systems generally do not automate this function, but they are easily constructed on graph paper. The following steps are used to construct such a plot:

1. The first step is to draw two horizontal lines on the graph paper. One is used to plot the vertical data and the other to plot the horizontal. These two lines show the location of each measurement point in inches with the outboard motor bearing being at zero.
2. Next, draw vertical lines that intersect the left-hand end of the two horizontal lines. These vertical lines form the amplitude scale for the two plots. Establish the amplitude scale based on the maximum energy level recorded in the broadband or narrowband windows.
3. The final step is to plot the actual measured amplitude at each measurement point on the machine-train. Start with the outboard motor bearing and move across the machine until the final data set is plotted.

Broadband Plots

The overall energy from the filtered broadband plotted against measurement location provides an approximation of the mode shape of the installed machine. Figure 14.8 illustrates a vertical broadband plot taken from a Spencer blower. Note that the motor appears to be flexing in the vertical direction. Extremely high amplitudes are present in the motor's outboard bearing and the amplitudes decrease at subsequent measurement points across the machine.

A mode curve exhibiting the shape shown in Figure 14.8 could indicate that the motor mountings, or the baseplate under the motor, are loose and that the motor is moving vertically. In fact, in the example from which this figure was taken, this is exactly what was happening. The blower's baseplate "floats" on a 1-in. thick cork pad, which

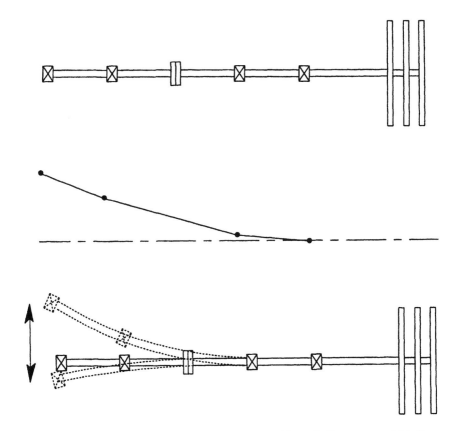

Figure 14.8 Vertical broadband mode shape for Spencer blower indicates potential failure.

is normally an acceptable practice. However, in this example, an inlet filter/silencer was mounted without support directly to the inlet located on the right end of the machine. The weight of the filter/silencer compressed the cork pad under the blower, which lifted the motor-end of the baseplate off of the cork pad. In this mode, the motor has complete freedom of movement in the vertical plane. In effect, it acts like a tuning fork and creates the high overall energy recorded on the mode plot.

Narrowband Plots

Narrowband plots permit the same type of evaluation for major vibration components such as fundamental running speed (1×) or gear mesh. The plots are constructed in the same way as for the broadbands, except that the amplitude values are for user-selected windows, or bands.

Using the previously mentioned example of the Spencer blower, Figure 14.9 is a plot of the fundamental (1×) frequency of the motor-blower shaft versus measurement location. Note that the vertical mode (see Figure 14.9) appears to be relatively normal,

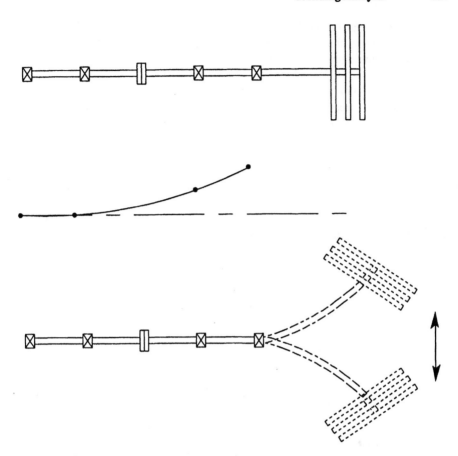

Figure 14.9 Horizontal narrowband (1×) mode shape indicates shaft deflection.

except for the motor looseness problem. The horizontal plot seems to indicate that the shaft is being severely deflected from its true centerline. In addition, the plot suggests that the deflection is outboard (i.e., toward the rotor) from the two blower-support bearings. This outboard deflection eliminates misalignment between the motor and blower as a possible source of the deflection. We must now determine what could cause the rotor to be deflected and why only in the horizontal direction.

The Spencer blower in this example provides air to a drying process in a metal-coating line. Its configuration includes an end-suction inlet that is in line with the shaft and a horizontal discharge that is perpendicular to the shaft. In this particular example, the source of the shaft deflection observed in the mode plot is aerodynamic instability.

The reason for this instability is that the blowers are incorrectly sized for the application and are running well outside their performance curve. In effect, the blowers have no back-pressure and are operating in a runout condition. The result of operating in this condition is that the design load intended to stabilize the rotor is no longer present. This causes the shaft to deflect or flex, generating the high amplitudes observed in the horizontal mode plot.

The problem is eliminated by restricting the discharge air flow from the blowers. By increasing the back-pressure, the blowers are able to operate within their normal envelope and the shaft deflection disappears.

Cross-Machine Comparisons

Cross-machine comparison is an extremely beneficial tool to the novice analyst. Most vibration-monitoring systems permit direct comparison of vibration data, both filtered window energy and complete signatures, acquired from two machines. This capability permits the analyst to compare a machine that is known to be in good operating condition directly with one that is perceived to have a problem. There are several ways that cross-machine comparisons can be made using microprocessor-based systems: multiple plots, ratio, and difference.

LIMITATIONS

Although quite valuable when used properly, trends do not allow the analyst to confirm that a problem exists or to determine the cause of incipient problems. Another limitation is the limited number of values the system can handle. Further, the data need to be normalized for speed, load, and process variables.

Number of Values

Some of the vibration-monitoring systems limit the number of data sets and duration of data that can be automatically trended. In most systems, the number of values that can be trended is limited to 8 to 12 data sets. Although this limitation prevents trending the machine over its useful life, it does not eliminate trending as a vibration-monitoring tool.

Data Normalization

Trend data that are not properly normalized for speed, load, and process variables are of little value. Because load and process-variable normalization require a little more time during the data-acquisition process, many programs do not perform these adjustments. If this is the case, it is best to discontinue the use of trends altogether.

As an example, Figure 14.10 illustrates the impact of load on vibration trends. The solid line represents the recorded raw broadband vibration levels. The dashed line is the same data adjusted for changes in load.

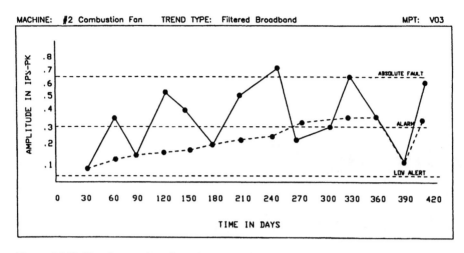

Figure 14.10 Trends must be adjusted or normalized for load changes.

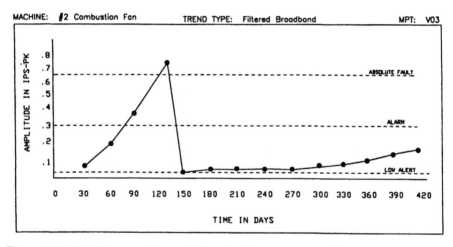

Figure 14.11 Baselines must be reset following repair.

Trend data also must be adjusted for maintenance and repair activities. Figure 14.11 illustrates an average trend curve that indicates a sharp rise in vibration levels. It also reflects that, after repair, the levels drop radically. At this point, all baseline and reference values should be reset. If this does not occur, the automatic trending capabilities of the computer-based system do not function properly.

Chapter 15

FAILURE-MODE ANALYSIS

All of the analysis techniques discussed to this point have been methods to determine if a potential problem exists within the machine-train or its associated systems. Failure-mode analysis is the next step required to pinpoint specifically the failure mode and identify which machine-train component is degrading.

Although failure-mode analysis identifies the number and symptoms of machine-train problems, it does not always identify the true root cause of problems. Root cause must be verified by visual inspection, additional testing, or other techniques such as operating dynamics analysis.

Failure-mode analysis is based on the assumption that certain failure modes are common to all machine-trains and all applications. It also assumes that the vibration patterns for each of these failure modes, when adjusted for process-system dynamics, are absolute and identifiable.

Two types of information are required to perform failure-mode analysis: (1) machine-train vibration signatures, both FFTs and time traces; and (2) practical knowledge of machine dynamics and failure modes. A number of failure-mode charts are available that describe the symptoms or abnormal vibration profiles that indicate potential problems exist. An example is the following description of the imbalance failure mode, which was obtained from a failure-mode chart: Single-plane imbalance generates a dominant fundamental (1x) frequency component with no harmonics (2x, 3x, etc.). Note, however, that the failure-mode charts are simplistic since many other machine-train problems also excite, or increase the amplitude of, the fundamental (1x) frequency component. In a normal vibration signature, 60 to 70% of the total overall, or broadband, energy is contained in the 1x frequency component. Any deviation from a state of equilibrium increases the energy level at this fundamental shaft speed.

COMMON GENERAL FAILURE MODES

Many of the common causes of failure in machinery components can be identified by understanding their relationship to the true running speed of the shaft within the machine-train.

Table 15.1 is a vibration troubleshooting chart that identifies some of the common failure modes. This table provides general guidelines for interpreting the most common abnormal vibration profiles. These guidelines, however, do not provide positive verification or identification of machine-train problems. Verification requires an understanding of the failure mode and how it appears in the vibration signature.

The sections that follow describe the most common machine-train failure modes: critical speeds, imbalance, mechanical looseness, misalignment, modulations, process instability, and resonance.

Critical Speeds

All machine-trains have one or more critical speeds that can cause severe vibration and damage to the machine. Critical speeds result from the phenomenon known as dynamic resonance.

Critical speed is a function of the natural frequency of dynamic components such as a rotor assembly, bearings, etc. All dynamic components have one or more natural frequencies that can be excited by an energy source that coincides with, or is in proximity to, that frequency. For example, a rotor assembly with a natural frequency of 1800 rpm cannot be rotated at speeds between 1782 and 1818 rpm without exciting the rotor's natural frequency.

Critical speed should not be confused with the mode shape of a rotating shaft. Deflection of the shaft from its true centerline (i.e., mode shape) elevates the vibration amplitude and generates dominant vibration frequencies at the rotor's fundamental and harmonics of the running speed. However, the amplitude of these frequency components tends to be much lower than those caused by operating at a critical speed of the rotor assembly. Also, the excessive vibration amplitude generated by operating at a critical speed disappears when the speed is changed. Vibrations caused by mode shape tend to remain through a much wider speed range or may even be independent of speed.

The unique natural frequencies of dynamic machine components are determined by the mass, freedom of movement, support stiffness, and other factors. These factors define the response characteristics of the rotor assembly (i.e., rotor dynamics) at various operating conditions.

Each critical speed has a well-defined vibration pattern. The first critical excites the fundamental (1x) frequency component; the second critical excites the secondary (2x) component; and the third critical excites the third (3x) frequency component.

Table 15.1 Vibration Troubleshooting Chart

Nature of Fault	Frequency of Dominant Vibration (Hz=rpm· 60)	Direction	Remarks
Rotating Members out of Balance	1 x rpm	Radial	A common cause of excess vibration in machinery
Misalignment & Bent Shaft	Usually 1 x rpm Often 2 x rpm Sometimes 3&4 x rpm	Radial & Axial	A common fault
Damaged Rolling Element Bearings (Ball, Roller, etc.)	Impact rates for the individual bearing components* Also vibrations at very high frequencies (20 to 60 kHz)	Radial & Axial	Uneven vibration levels, often with shocks. *Impact-Rates:
Journal Bearings Loose in Housings	Sub-harmonics of shaft rpm, exactly 1/2 or 1/3 x rpm	Primarily Radial	Looseness may only develop at operating speed and temperature (eg. turbomachines).
Oil Film Whirl or Whip in Journal Bearings	Slightly less than half shaft speed (42% to 48%)	Primarily Radial	Applicable to high-speed (eg. turbo) machines.
Hysteresis Whirl	Shaft critical speed	Primarily Radial	Vibrations excited when passing through critical shaft speed are maintained at higher shaft speeds. Can sometimes be cured by checking tightness of rotor components
Damaged or worn gears	Tooth meshing frequencies (shaft rpm x number of teeth) and harmonics	Radial & Axial	Sidebands around tooth meshing frequencies indicate modulation (eg. eccentricity) at frequency corresponding to sideband spacings. Normally only detectable with very narrow-band analysis.
Mechanical Looseness	2 x rpm		
Faulty Belt Drive	1, 2, 3 & 4 x rpm of belt	Radial	
Unbalanced Reciprocating Forces and Couples	1 x rpm and/or multiples for higher order unbalance	Primarily Radial	
Increased Turbulence	Blade & Vane passing frequencies and harmonics	Radial & Axial	Increasing levels indicate increasing turbulence.
Electrically Induced Vibrations	1 x rpm or 1 or 2 times sychronous frequency	Radial & Axial	Should disappear when turning off the power

Source: Predictive Maintenance for Process Machinery, R. Keith Mobley, Technology for Energy Corp., 1988.

The best way to confirm a critical-speed problem is to change the operating speed of the machine-train. If the machine is operating at a critical speed, the amplitude of the vibration components (1×, 2×, or 3×) will immediately drop when the speed is changed. If the amplitude remains relatively constant when the speed is changed, the problem is not critical speed.

Imbalance

The term *balance* means that all forces generated by, or acting on, the rotating element of a machine-train are in a state of equilibrium. Any change in this state of equilibrium creates an imbalance. In the global sense, imbalance is one of the most common abnormal vibration profiles exhibited by all process machinery.

Theoretically, a perfectly balanced machine that has no friction in the bearings would experience no vibration and would have a perfect vibration profile—a perfectly flat, horizontal line. However, there are no perfectly balanced machines in existence. All machine-trains exhibit some level of imbalance, which has a dominant frequency component at the fundamental running speed (1×) of each shaft.

An imbalance profile can be excited due to the combined factors of mechanical imbalance, lift/gravity differential effects, aerodynamic and hydraulic instabilities, process loading, and, in fact, all failure modes.

Mechanical Imbalance

It is incorrect to assume that mechanical imbalance must exist to create an imbalance condition within the machine. Mechanical imbalance, however, is the only form of imbalance that is corrected by balancing the rotating element. When all failures are considered, the number of machine problems that are the result of actual mechanical rotor imbalance is relatively small.

Single-Plane Mechanical Imbalance

Single-plane mechanical imbalance excites the fundamental (1×) frequency component, which is typically the dominant amplitude in a signature. Because there is only one point of imbalance, only one high spot occurs as the rotor completes each revolution. The vibration signature also may contain lower level frequencies reflecting bearing defects and passing frequencies. Figure 15.1 illustrates single-plane imbalance.

Because mechanical imbalance is multidirectional, it appears in both the vertical and horizontal directions at the machine's bearing pedestals. The actual amplitude of the 1× component generally is not identical in the vertical and horizontal directions and both generally contain elevated vibration levels at 1×.

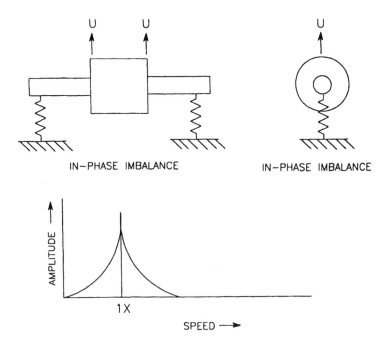

Figure 15.1 Single-plane imbalance.

The difference between the vertical and horizontal values is a function of the bearing-pedestal stiffness. In most cases, the horizontal plane has a greater freedom of movement and, therefore, contains higher amplitudes at 1x than the vertical plane.

Multiplane Mechanical Imbalance

Multiplane mechanical imbalance generates multiple harmonics of running speed. The actual number of harmonics depends on the number of imbalance points, the severity of imbalance, and the phase angle between imbalance points.

Figure 15.2 illustrates a case of multiplane imbalance in which there are four out-of-phase imbalance points. The resultant vibration profile contains dominant frequencies at 1x, 2x, 3x, and 4x. The actual amplitude of each of these components is determined by the amount of imbalance at each of the four points, but the 1x component should always be higher than any subsequent harmonics.

Lift/Gravity Differential Imbalance

Lift, which is designed into a machine-train's rotating elements to compensate for the effects of gravity acting on the rotor, is another source of imbalance. Because lift does not always equal gravity, there is always some imbalance in machine-trains. The vibration component due to the lift/gravity differential effect appears at the fundamental or 1x frequency.

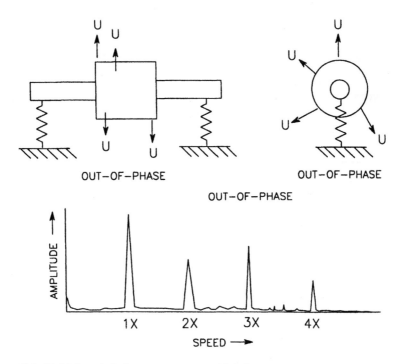

Figure 15.2 Multiplane imbalance generates multiple harmonics.

Other

In fact, all failure modes create some form of imbalance in a machine, as do aerodynamic instability, hydraulic instability, and process loading. The process loading of most machine-trains varies, at least slightly, during normal operations. These vibration components appear at the 1× frequency.

Mechanical Looseness

Looseness, which can be present in both the vertical and horizontal planes, can create a variety of patterns in a vibration signature. In some cases, the fundamental (1×) frequency is excited. In others, a frequency component at one-half multiples of the shaft's running speed (0.5×, 1.5×, 2.5×, etc.) is present. In almost all cases, there are multiple harmonics, both full and half.

Vertical Looseness

Mechanical looseness in the vertical plane generates a series of harmonic and half-harmonic frequency components. Figure 15.3 is a simple example of a vertical mechanical looseness signature.

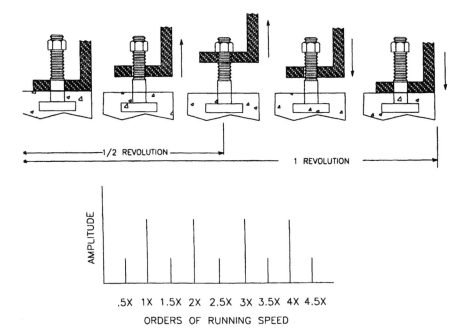

Figure 15.3 Vertical mechanical looseness has a unique vibration profile.

In most cases, the half-harmonic components are about one-half of the amplitude of the harmonic components. They result from the machine-train lifting until stopped by the bolts. The impact as the machine reaches the upper limit of travel generates a frequency component at one-half multiples (i.e., orders) of running speed. As the machine returns to the bottom of its movement, its original position, a larger impact occurs that generates the full harmonics of running speed.

The difference in amplitude between the full harmonics and half-harmonics is caused by the effects of gravity. As the machine lifts to its limit of travel, gravity resists the lifting force. Therefore, the impact force that is generated as the machine foot contacts the mounting bolt is the difference between the lifting force and gravity. As the machine drops, the force of gravity combines with the force generated by imbalance. The impact force as the machine foot contacts the foundation is the sum of the force of gravity and the force resulting from imbalance.

Horizontal Looseness

Figure 15.4 illustrates horizontal mechanical looseness, which is also common to machine-trains. In this example, the machine's support legs flex in the horizontal plane. Unlike the vertical looseness illustrated in Figure 15.3, gravity is uniform at each leg and there is no increased impact energy as the leg's direction is reversed.

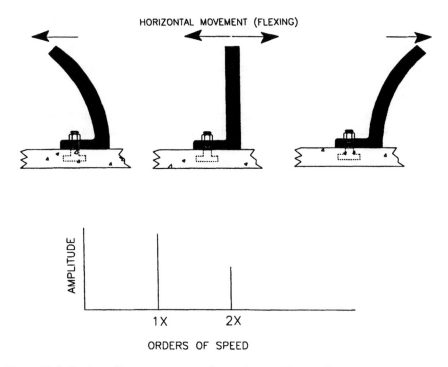

Figure 15.4 Horizontal looseness creates first and second harmonics.

Horizontal mechanical looseness generates a combination of first (1×) and second (2×) harmonic vibrations. Since the energy source is the machine's rotating shaft, the timing of the flex is equal to one complete revolution of the shaft, or 1×. During this single rotation, the mounting legs flex to their maximum deflection on both sides of neutral. The double change in direction as the leg first deflects to one side then the other generates a frequency at two times (2×) the shaft's rotating speed.

Other

There are a multitude of other forms of mechanical looseness (besides vertical and horizontal movement of machine legs) that are typical for manufacturing and process machinery. Most forms of pure mechanical looseness result in an increase in the vibration amplitude at the fundamental (1×) shaft speed. In addition, looseness generates one or more harmonics (i.e., 2×, 3×, 4×, or combinations of harmonics and half-harmonics).

However, not all looseness generates this classic profile. For example, excessive bearing and gear clearances do not generate multiple harmonics. In these cases, the vibration profile contains unique frequencies that indicate looseness, but the profile varies depending on the nature and severity of the problem.

With sleeve or Babbitt bearings, looseness is displayed as an increase in subharmonic frequencies (i.e., less than the actual shaft speed, such as 0.5×). Rolling-element bearings display elevated frequencies at one or more of their rotational frequencies. Excessive gear clearance increases the amplitude at the gear-mesh frequency and its sidebands.

Other forms of mechanical looseness increase the noise floor across the entire bandwidth of the vibration signature. While the signature does not contain a distinct peak or series of peaks, the overall energy contained in the vibration signature is increased. Unfortunately, the increase in noise floor cannot always be used to detect mechanical looseness. Some vibration instruments lack sufficient dynamic range to detect changes in the signature's noise floor.

Misalignment

This condition is virtually always present in machine-trains. Generally, we assume that misalignment exists between shafts that are connected by a coupling, V-belt, or other intermediate drive. However, it can exist between bearings of a solid shaft and at other points within the machine.

How misalignment appears in the vibration signature depends on the type of misalignment. Figure 15.5 illustrates three types of misalignment (i.e., internal, offset, and angular). These three types excite the fundamental (1×) frequency component because they create an apparent imbalance condition in the machine.

Internal (i.e., bearing) and offset misalignment also excite the second (2×) harmonic frequency. Two high spots are created by the shaft as it turns though one complete revolution. These two high spots create the first (1×) and second harmonic (2×) components.

Angular misalignment can take several signature forms and excites the fundamental (1×) and secondary (2×) components. It can excite the third (3×) harmonic frequency depending on the actual phase relationship of the angular misalignment. It also creates a strong axial vibration.

Modulations

Modulations are frequency components that appear in a vibration signature, but cannot be attributed to any specific physical cause, or forcing function. Although these frequencies are, in fact, ghosts or artificial frequencies, they can result in significant damage to a machine-train. The presence of ghosts in a vibration signature often leads to misinterpretation of the data.

Ghosts are caused when two or more frequency components couple, or merge, to form another discrete frequency component in the vibration signature. This generally occurs with multiple-speed machines or a group of single-speed machines.

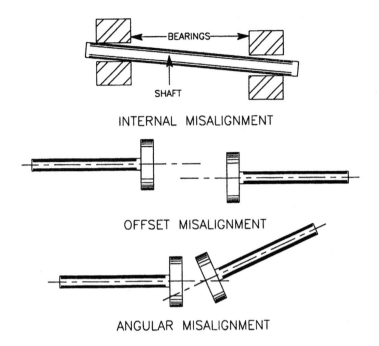

Figure 15.5 Three types of misalignment.

Note that the presence of modulation, or ghost peaks, is not an absolute indication of a problem within the machine-train. Couple effects may simply increase the amplitude of the fundamental running speed and do little damage to the machine-train. However, this increased amplitude will amplify any defects within the machine-train.

Coupling can have an additive effect on the modulation frequencies, as well as being reflected as a differential or multiplicative effect. These concepts are discussed in the sections that follow.

Take as an example the case of a 10-tooth pinion gear turning at 10 rpm while driving a 20-tooth bullgear having an output speed of 5 rpm. This gear set generates real frequencies at 5, 10, and 100 rpm (i.e., 10 teeth × 10 rpm). This same set also can generate a series of frequencies (i.e., sum and product modulations) at 15 rpm (i.e., 10 rpm + 5 rpm) and 150 rpm (i.e., 15 rpm × 10 teeth). In this example, the 10-rpm input speed coupled with the 5-rpm output speed to create ghost frequencies driven by this artificial fundamental speed (15 rpm).

Sum Modulation

This type of modulation, which is described in the preceding example, generates a series of frequencies that includes the fundamental shaft speeds, both input and output, and fundamental gear-mesh profile. The only difference between the real frequencies and the ghost is their location on the frequency scale. Instead of being at the

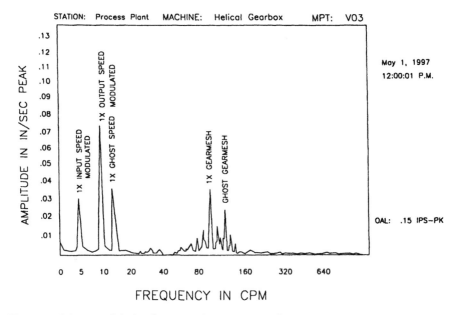

STATION: Process Plant   MACHINE: Helical Gearbox   MPT: VO3

May 1, 1997
12:00:01 P.M.

OAL: .15 IPS-PK

Figure 15.6 Sum modulation for a speed-increaser gearbox.

actual shaft-speed frequency, the ghost appears at frequencies equal to the sum of the input and output shaft speeds. Figure 15.6 illustrates this for a speed-increaser gearbox.

Difference Modulation

In this case, the resultant ghost, or modulation, frequencies are generated by the difference between two or more speeds (see Figure 15.7). If we use the same example as before, the resultant ghost frequencies appear at 5 rpm (i.e., 10 rpm − 5 rpm) and 50 rpm (i.e., 5 rpm × 10 teeth). Note that the 5-rpm couple frequency coincides with the real output speed of 5 rpm. This results in a dramatic increase in the amplitude of one real running-speed component and the addition of a false gear-mesh peak.

This type of coupling effect is common in single-reduction/increase gearboxes or other machine-train components where multiple running or rotational speeds are relatively close together or even integer multiples of one another. It is more destructive than other forms of coupling in that it coincides with real vibration components and tends to amplify any defects within the machine-train.

Product Modulation

With product modulation, the two speeds couple in a multiplicative manner to create a set of artificial frequency components (see Figure 15.8). In the previous example, product modulations occur at 50 rpm (i.e., 10 rpm × 5 rpm) and 500 rpm (i.e., 50 rpm × 10 teeth).

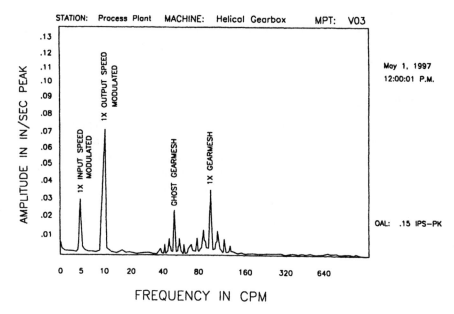

Figure 15.7 Difference modulation for a speed-increaser gearbox.

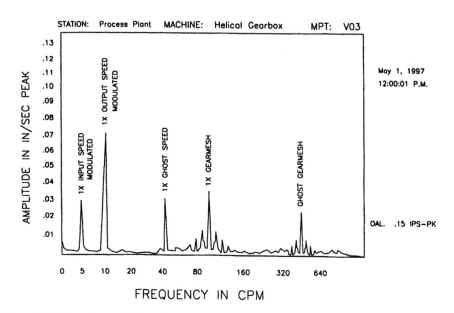

Figure 15.8 Product modulation for a speed-increaser gearbox.

Beware that this type of coupling often may go undetected in a normal vibration analysis. Since the ghost frequencies are relatively high compared to the expected real frequencies, they are often outside the monitored frequency range used for data acquisition and analysis.

Process Instability

Normally associated with bladed or vaned machinery such as fans and pumps, process instability creates an imbalanced condition within the machine. In most cases, it excites the fundamental (1x) and blade-pass/vane-pass frequency components. Unlike true mechanical imbalance, the blade-pass and vane-pass frequency components are broader and have more energy in the form of sideband frequencies.

In most cases, this failure mode also excites the third (3x) harmonic frequency and creates strong axial vibration. Depending on the severity of the instability and the design of the machine, process instability also can create a variety of shaft-mode shapes. In turn, this excites the 1x, 2x, and 3x radial vibration components.

Resonance

Resonance is defined as a large-amplitude vibration caused by a small periodic stimulus having the same, or nearly the same, period as the system's natural vibration. In other words, an energy source with the same, or nearly the same, frequency as the natural frequency of a machine-train or structure will excite that natural frequency. The result is a substantial increase in the amplitude of the natural frequency component.

The key point to remember is that a very low amplitude energy source can cause massive amplitudes when its frequency coincides with the natural frequency of a machine or structure. Higher levels of input energy can cause catastrophic, near instantaneous failure of the machine or structure.

Every machine-train has one or more natural frequencies. If one of these frequencies is excited by some component of the normal operation of the system, the machine structure will amplify the energy, which can cause severe damage.

An example of resonance is a tuning fork. If you activate a tuning fork by striking it sharply, the fork vibrates rapidly. As long as it is held suspended, the vibration decays with time. However, if you place it on a desk top, the fork could potentially excite the natural frequency of the desk, which would dramatically amplify the vibration energy.

The same thing can occur if one or more of the running speeds of a machine excites the natural frequency of the machine or its support structure. Resonance is a very destructive vibration and, in most cases, it will cause major damage to the machine or support structure.

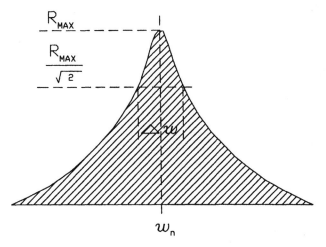

Figure 15.9 Resonance response.

There are two major classifications of resonance found in most manufacturing and process plants: static and dynamic. Both types exhibit a broad-based, high-amplitude frequency component when viewed in a FFT vibration signature. Unlike meshing or passing frequencies, the resonance frequency component does not have modulations or sidebands. Instead, resonance is displayed as a single, clearly defined peak.

As illustrated in Figure 15.9, a resonance peak represents a large amount of energy. This energy is the result of both the amplitude of the peak and the broad area under the peak. This combination of high peak amplitude and broad-based energy content is typical of most resonance problems. The damping system associated with a resonance frequency is indicated by the sharpness or width of the response curve, ω_n, when measured at the half-power point. R_{MAX} is the maximum resonance and $R_{MAX}/\sqrt{2}$ is the half-power point for a typical resonance-response curve.

Static Resonance

When the natural frequency of a stationary, or nondynamic, structure is energized, it will resonate. This type of resonance is classified as static resonance and is considered to be a nondynamic phenomenon. Nondynamic structures in a machine-train include casings, bearing-support pedestals, and structural members such as beams, piping, etc.

Since static resonance is a nondynamic phenomenon, it is generally not associated with the primary running speed of any associated machinery. Rather, the source of static resonance can be any energy source that coincides with the natural frequency of any stationary component. For example, an I-beam support on a continuous annealing line may be energized by the running speed of a roll. However, it also can be made to

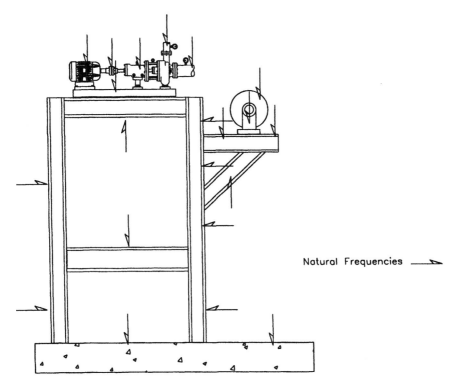

Figure 15.10 Typical discrete natural frequency locations in structural members.

resonate by a bearing frequency, overhead crane, or any of a multitude of other energy sources.

The actual resonant frequency depends on the mass, stiffness, and span of the excited member. In general terms, the natural frequency of a structural member is inversely proportional to the mass and stiffness of the member. In other words, a large turbocompressor's casing will have a lower natural frequency than that of a small end-suction centrifugal pump.

Figure 15.10 illustrates a typical structural-support system. The natural frequencies of all support structures, piping, and other components are functions of mass, span, and stiffness. Each of the arrows on Figure 15.10 indicates a structural member or stationary machine component having a unique natural frequency. Note that each time a structural span is broken or attached to another structure, the stiffness changes. As a result, the natural frequency of that segment also changes.

While most stationary machine components move during normal operation, they are not always resonant. Some degree of flexing or movement is common in most stationary machine-trains and structural members. The amount of movement depends on the

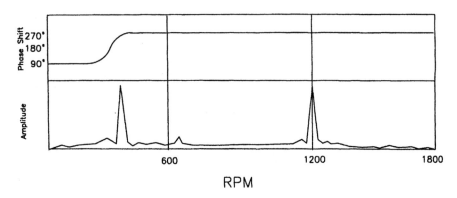

Figure 15.11 Dynamic resonance phase shift.

spring constant or stiffness of the member. However, when an energy source coincides and couples with the natural frequency of a structure, excessive and extremely destructive vibration amplitudes result.

Dynamic Resonance

When the natural frequency of a rotating, or dynamic, structure (e.g., rotor assembly in a fan) is energized, the rotating element resonates. This phenomenon is classified as dynamic resonance and the rotor speed at which it occurs is referred to as the critical.

In most cases, dynamic resonance appears at the fundamental running speed or one of the harmonics of the excited rotating element. However, it also can occur at other frequencies. As in the case of static resonance, the actual natural frequencies of dynamic members depend on the mass, bearing span, shaft and bearing-support stiffness, and a number of other factors.

Confirmation Analysis

In most cases, the occurrence of dynamic resonance can be quickly confirmed. When monitoring phase and amplitude, resonance is indicated by a 180-degree phase shift as the rotor passes through the resonant zone. Figure 15.11 illustrates a dynamic resonance at 500 rpm, which shows a dramatic amplitude increase in the frequency-domain display. This is confirmed by the 180-degree phase shift in the time-domain plot. Note that the peak at 1200 rpm is not resonance. The absence of a phase shift, coupled with the apparent modulations in the FFT, discount the possibility that this peak is resonance related.

Common Confusions

Vibration analysts often confuse resonance with other failure modes. Because many of the common failure modes tend to create abnormally high vibration levels that appear to be related to a speed change, analysts tend to miss the root cause of these problems.

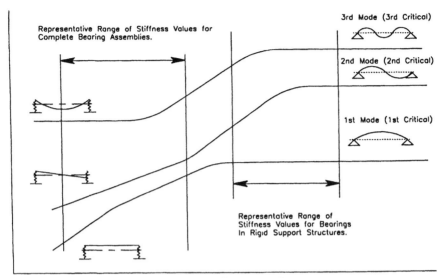

Figure 15.12 Dynamic resonance plot.

Dynamic resonance generates abnormal vibration profiles that tend to coincide with the fundamental (1x) running speed, or one or more of the harmonics, of a machine-train. This often leads the analyst to incorrectly diagnose the problem as imbalance or misalignment. The major difference is that dynamic resonance is the result of a relatively small energy source, such as the fundamental running speed, that results in a massive amplification of the natural frequency of the rotating element.

Function of Speed
The high amplitudes at the rotor's natural frequency are strictly speed dependent. If the energy source, in this case speed, changes to a frequency outside the resonant zone, the abnormal vibration will disappear.

In most cases, running speed is the forcing function that excites the natural frequency of the dynamic component. As a result, rotating equipment is designed to operate at primary rotor speeds that do not coincide with the rotor assembly's natural frequencies. Most low- to moderate-speed machines are designed to operate below the first critical speed of the rotor assembly.

Higher speed machines may be designed to operate between the first and second, or second and third, critical speeds of the rotor assembly. As these machines accelerate through the resonant zones or critical speeds, their natural frequency is momentarily excited. As long as the ramp rate limits the duration of excitation, this mode of operation is acceptable. However, care must be taken to ensure that the transient time through the resonant zone is as short as possible.

Figure 15.12 illustrates a typical critical-speed or dynamic-resonance plot. This figure is a plot of the relationship between rotor-support stiffness (X-axis) and critical rotor speed (Y-axis). Rotor-support stiffness depends on the geometry of the rotating element (i.e., shaft and rotor) and the bearing-support structure. These are the two dominant factors that determine the response characteristics of the rotor assembly.

FAILURE MODES BY MACHINE-TRAIN COMPONENT

In addition to identifying general failure modes that are common to many types of machine-train components, failure-mode analysis can be used to identify failure modes for specific components in a machine-train. However, care must be exercised when analyzing vibration profiles, because the data may reflect induced problems. Induced problems affect the performance of a specific component, but are not caused by that component. For example, an abnormal outer-race passing frequency may indicate a defective rolling-element bearing. It also can indicate that abnormal loading caused by misalignment, roll bending, process instability, etc., has changed the load zone within the bearing. In the latter case, replacing the bearing does not resolve the problem and the abnormal profile will still be present after the bearing is changed.

Bearings: Rolling Element

Bearing defects are one of the most common faults identified by vibration monitoring programs. Although bearings do wear out and fail, these defects are normally symptoms of other problems within the machine-train or process system. Therefore, extreme care must be exercised to ensure that the real problem is identified, not just the symptom. In a rolling-element, or antifriction, bearing vibration profile, three distinct sets of frequencies can be found: natural, rotational, and defect.

Natural Frequency

Natural frequencies are generated by impacts of the internal parts of a rolling-element bearing. These impacts are normally the result of slight variations in load and imperfections in the machined bearing surfaces. As their name implies, these are natural frequencies and are present in a new bearing that is in perfect operating condition.

The natural frequencies of rolling-element bearings are normally well above the maximum frequency range, F_{MAX}, used for routine machine-train monitoring. As a result, they are rarely observed by predictive maintenance analysts. Generally, these frequencies are between 20 kHz and 1 MHz. Therefore, some vibration-monitoring programs use special high-frequency or ultrasonic monitoring techniques such as high-frequency domain (HFD).

Note, however, that little is gained from monitoring natural frequencies. Even in cases of severe bearing damage, these high-frequency components add little to the analyst's ability to detect and isolate bad bearings.

Rotational Frequency

Four normal rotational frequencies are associated with rolling-element bearings: fundamental train frequency (FTF), ball/roller spin, ball-pass outer-race, and ball-pass inner-race. The following are definitions of abbreviations that are used in the discussion that follows:

BD = Ball or roller diameter
PD = Pitch diameter
β = Contact angle (for roller = 0)
n = Number of balls or rollers
f_r = Relative speed between the inner and outer race (rps).

Fundamental Train Frequency

The bearing cage generates the FTF as it rotates around the bearing races. The cage properly spaces the balls or rollers within the bearing races, in effect, by tying the rolling elements together and providing uniform support. Some friction exists between the rolling elements and the bearing races, even with perfect lubrication. This friction is transmitted to the cage, which causes it to rotate around the bearing races.

Because this is a friction-driven motion, the cage turns much slower than the inner race of the bearing. Generally, the rate of rotation is slightly less than one-half of the shaft speed. The FTF is calculated by the following equation:

$$FTF = \frac{1}{2}f_r\left[1 - \frac{BD}{PD}\right]$$

Ball-Spin Frequency

Each of the balls or rollers within a bearing rotates around its own axis as it rolls around the bearing races. This spinning motion is referred to as ball spin, which generates a ball-spin frequency (BSF) in a vibration signature. The speed of rotation is determined by the geometry of the bearing (i.e., diameter of the ball or roller, and bearing races) and is calculated by:

$$BSF = \frac{1}{2}\frac{PD}{BD}xf_r\left[1 - \left(\frac{BD}{PD}\right)^2 x\cos^2\beta\right]$$

Ball-Pass Outer-Race

The ball or rollers passing the outer race generate the ball-pass outer-race frequency (BPFO), which is calculated by:

$$BPFO = \frac{n}{2} \times f_r\left(1 - \frac{BD}{PD}x\cos\beta\right)$$

Ball-Pass Inner-Race

The speed of the ball/roller rotating relative to the inner race generates the ball-pass inner-race rotational frequency (BPFI). The inner race rotates at the same speed as the shaft and the complete set of balls/rollers passes at a slower speed. They generate a passing frequency that is determined by:

$$BPFI = \frac{n}{2} \times f_r \left(1 + \frac{BD}{PD} x \cos \beta \right)$$

Defect Frequencies

Rolling-element bearing defect frequencies are the same as their rotational frequencies, except for the BSF. If there is a defect on the inner race, the BPFI amplitude increases because the balls or rollers contact the defect as they rotate around the bearing. The BPFO is excited by defects in the outer race.

When one or more of the balls or rollers have a defect such as a spall (i.e., a missing chip of material), the defect impacts both the inner and outer race each time one revolution of the rolling element is made. Therefore, the defect vibration frequency is visible at two times (2×) the BSF rather than at its fundamental (1×) frequency.

Bearings: Sleeve (Babbitt)

In normal operation, a sleeve bearing provides a uniform oil film around the supported shaft. Because the shaft is centered in the bearing, all forces generated by the rotating shaft, and all forces acting on the shaft, are equal. Figure 15.13 shows the balanced forces on a normal bearing.

Lubricating-film instability is the dominant failure mode for sleeve bearings. This instability is typically caused by eccentric, or off-center, rotation of the machine shaft resulting from imbalance, misalignment, or other machine or process-related problems. Figure 15.14 shows a Babbitt bearing that exhibits instability.

When oil-film instability or oil whirl occurs, frequency components at fractions (i.e., 1/4, 1/3, 3/8, etc.) of the fundamental (1×) shaft speed are excited. As the severity of the instability increases, the frequency components become more dominant in a band between 0.40 and 0.48 of the fundamental (1×) shaft speed. When the instability becomes severe enough to isolate within this band, it is called oil whip. Figure 15.15 shows the effect of increased velocity on a Babbitt bearing.

Chains and Sprockets

Chain drives function in essentially the same basic manner as belt drives. However, instead of tension, chains depend on the mechanical meshing of sprocket teeth with the chain links.

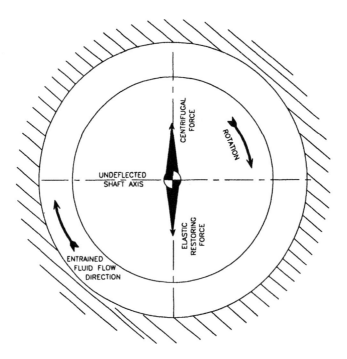

Figure 15.13 A normal Babbitt bearing has balanced forces.

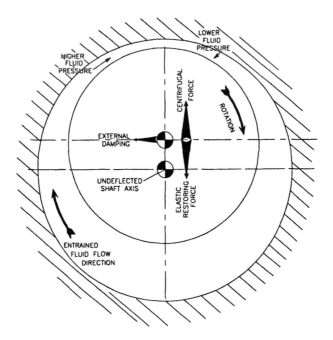

Figure 15.14 Dynamics of Babbitt bearing that exhibits instability.

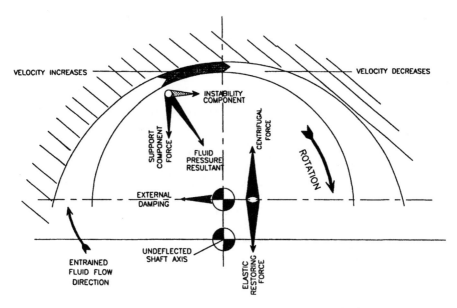

Figure 15.15 Increased velocity generates an unbalanced force in a Babbitt bearing.

Gears

All gear sets create a frequency component referred to as gear mesh. The fundamental gear-mesh frequency is equal to the number of gear teeth times the running speed of the shaft. In addition, all gear sets create a series of sidebands or modulations that are visible on both sides of the primary gear-mesh frequency.

Normal Profile

In a normal gear set, each of the sidebands is spaced by exactly the 1× running speed of the input shaft and the entire gear mesh is symmetrical as seen in Figure 15.16. In addition, the sidebands always occur in pairs, one below and one above the gear-mesh frequency, and the amplitude of each pair is identical (Figure 15.17).

If we split the gear-mesh profile for a normal gear by drawing a vertical line through the actual mesh (i.e., number of teeth times the input shaft speed), the two halves would be identical. Therefore, any deviation from a symmetrical profile indicates a gear problem. However, care must be exercised to ensure that the problem is internal to the gears and not induced by outside influences.

External misalignment, abnormal induced loads, and a variety of other outside influences destroy the symmetry of a gear-mesh profile. For example, a single-reduction gearbox used to transmit power to a mold-oscillator system on a continuous caster drives two eccentric cams. The eccentric rotation of these two cams is transmitted

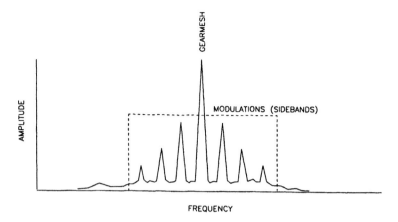

Figure 15.16 Normal gear set profile is symmetrical.

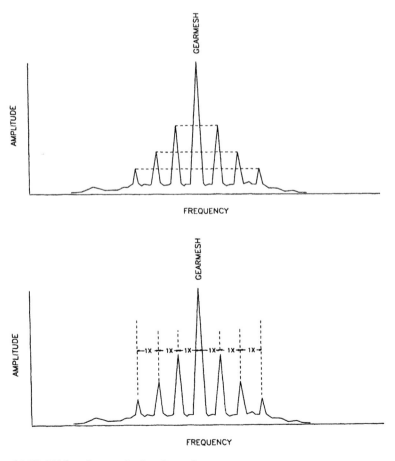

Figure 15.17 Sidebands are paired and equal.

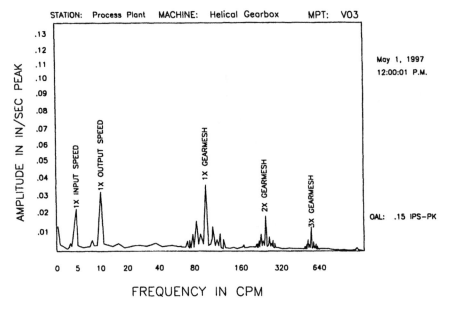

Figure 15.18 Typical defective gear-mesh signature.

directly into the gearbox, creating the appearance of eccentric meshing of the gears. However, this abnormal induced load actually destroys the spacing and amplitude of the gear-mesh profile.

Defective Gear Profiles

If the gear set develops problems, the amplitude of the gear-mesh frequency increases and the symmetry of the sidebands changes. The pattern illustrated in Figure 15.18 is typical of a defective gear set, where OAL is the broadband, or total, energy. Note the asymmetrical relationship of the sidebands.

Excessive Wear

Figure 15.19 is the vibration profile of a worn gear set. Note that the spacing between the sidebands is erratic and is no longer evenly spaced by the input shaft speed frequency. The sidebands for a worn gear set tend to occur between the input and output speeds and are not evenly spaced.

Cracked or Broken Teeth

Figure 15.20 illustrates the profile of a gear set with a broken tooth. As the gear rotates, the space left by the chipped or broken tooth increases the mechanical clearance between the pinion and bullgear. The result is a low-amplitude sideband to the left of the actual gear-mesh frequency. When the next (i.e., undamaged) teeth mesh, the added clearance results in a higher energy impact. The sideband to the right of the mesh frequency has a much higher amplitude. As a result, the paired sidebands have a

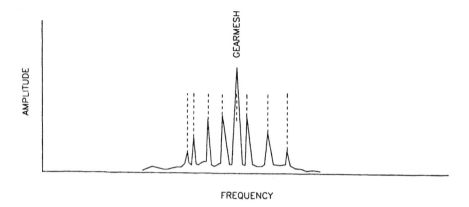

Figure 15.19 Wear or excessive clearance changes the sideband spacing.

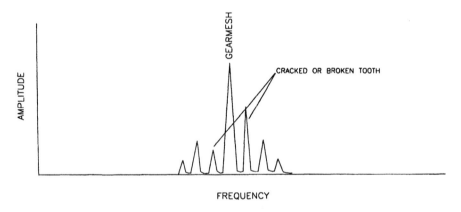

Figure 15.20 A broken tooth will produce an asymmetrical sideband profile.

nonsymmetrical amplitude, which is due to the disproportional clearance and impact energy.

Improper Shaft Spacing

In addition to gear-tooth wear, variations in the center-to-center distance between shafts create erratic spacing and amplitude in a vibration signature. If the shafts are too close together, the sideband spacing tends to be at input shaft speed, but the amplitude is significantly reduced. This condition causes the gears to be deeply meshed (i.e., below the normal pitch line), so the teeth maintain contact through the entire mesh. This loss of clearance results in lower amplitudes, but it exaggerates any tooth-profile defects that may be present.

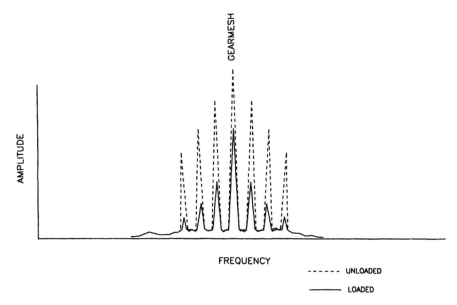

Figure 15.21 Unloaded gear has much higher vibration levels.

If the shafts are too far apart, the teeth mesh above the pitch line, which increases the clearance between teeth and amplifies the energy of the actual gear-mesh frequency and all of its sidebands. In addition, the load-bearing characteristics of the gear teeth are greatly reduced. Because the force is focused on the tip of each tooth where there is less cross-section, the stress in each tooth is greatly increased. The potential for tooth failure increases in direct proportion to the amount of excess clearance between the shafts.

Load Changes

The energy and vibration profiles of gear sets change with load. When the gear is fully loaded, the profiles exhibit the amplitudes discussed previously. When the gear is unloaded, the same profiles are present, but the amplitude increases dramatically. The reason for this change is gear-tooth roughness. In normal practice, the backside of the gear tooth is not finished to the same smoothness as the power, or drive, side. Therefore, there is more looseness on the nonpower, or back, side of the gear. Figure 15.21 illustrates the relative change between a loaded and unloaded gear profile.

Jackshafts and Spindles

Another form of intermediate drive consists of a shaft with some form of universal connection on each end that directly links the prime mover to a driven unit (see Figures 15.22 and 15.23). Jackshafts and spindles are typically used in applications where the driver and driven unit are misaligned.

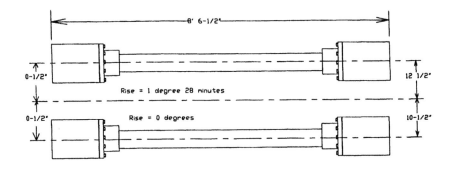

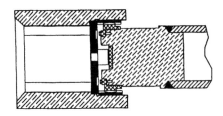

Figure 15.22 Typical gear-type spindle.

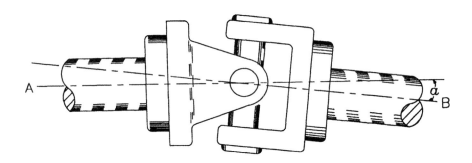

Figure 15.23 Typical universal-type jackshaft.

Most of the failure modes associated with jackshafts and spindles are the result of lubrication problems or fatigue failure resulting from overloading. However, the actual failure mode generally depends on the configuration of the flexible drive.

Lubrication Problems

Proper lubrication is essential for all jackshafts and spindles. A critical failure point for spindles (see Figure 15.22) is in the mounting pod that provides the connection between the driver and driven machine components. Mounting pods generally use either a spade-and-slipper or a splined mechanical connector. In both cases, regular application of a suitable grease is essential for prolonged operation. Without proper lubrication, the mating points between the spindle's mounting pod and the machine-train components impact each time the torsional power varies between the primary driver and driven component of the machine-train. The resulting mechanical damage can cause these critical drive components to fail.

In universal-type jackshafts like the one illustrated in Figure 15.23, improper lubrication results in nonuniform power transmission. The absence of a uniform grease film causes the pivot points within the universal joints to bind and restrict smooth power transmission.

The typical result of poor lubrication, which results in an increase in mechanical looseness, is an increase of those vibration frequencies associated with the rotational speed. In the case of gear-type spindles (Figure 15.22), there will be an increase in both the fundamental (1×) and second harmonic (2×). Because the resulting forces generated by the spindle are similar to angular misalignment, there also will be a marked increase in the axial energy generated by the spindle.

The universal-coupling configuration used by jackshafts (Figure 15.23) generates an elevated vibration frequency at the fourth (4×) harmonic of its true rotational speed. This failure mode is caused by the binding that occurs as the double pivot points move through a complete rotation.

Fatigue

Spindles and jackshafts are designed to transmit torsional power between a driver and driven unit that are not in the same plane or that have a radical variation in torsional power. Typically, both conditions are present when these flexible drives are used.

Both the jackshaft and spindle are designed to absorb transient increases or decreases in torsional power caused by twisting. In effect, the shaft or tube used in these designs winds, much like a spring, as the torsional power increases. Normally, this torque and the resultant twist of the spindle are maintained until the torsional load is reduced. At that point, the spindle unwinds, releasing the stored energy that was generated by the initial transient.

Repeated twisting of the spindle's tube or the solid shaft used in jackshafts results in a reduction in the flexible drive's stiffness. When this occurs, the drive loses some of its ability to absorb torsional transients. As a result, damage may result to the driven unit.

Unfortunately, the limits of single-channel, frequency-domain data acquisition prevent accurate measurement of this failure mode. Most of the abnormal vibration that results from fatigue occurs in the relatively brief time interval associated with startup, when radical speed changes occur, or during shutdown of the machine-train. As a result, this type of data acquisition and analysis cannot adequately capture these tran-

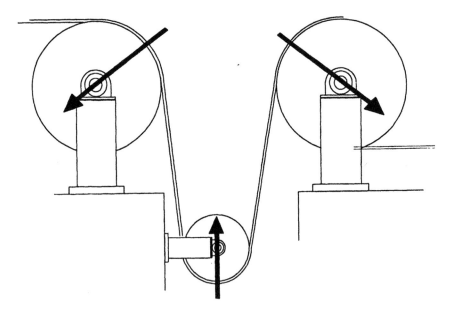

Figure 15.24 Load zones determined by wrap.

sients. However, the loss of stiffness caused by fatigue increases the apparent
mechanical looseness observed in the steady-state, frequency-domain vibration signature. In most cases, this is similar to the mechanical looseness.

Process Rolls

Process rolls commonly encounter problems or fail due to being subjected to induced
(variable) loads and from misalignment.

Induced (Variable) Loads

Process rolls are subjected to variable loads that are induced by strip tension, tracking,
and other process variables. In most cases, these loads are directional. They not only
influence the vibration profile, but determine the location and orientation of data
acquisition.

Strip Tension or Wrap

Figure 15.24 illustrates the wrap of the strip as it passes over a series of rolls in a continuous-process line. The orientation and contact area of this wrap determines the load
zone on each roll. In this example, the strip wrap is limited to one-quarter of the roll
circumference. The load zone, or vector, on the two top rolls is on a 45-degree angle
to the passline. Therefore, the best location for the primary radial measurement is at
45 degrees opposite to the load vector. The secondary radial measurement should be
90 degrees to the primary. On the top-left roll, the secondary measurement point

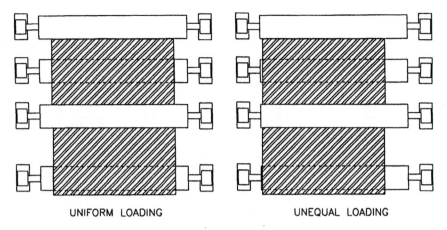

UNIFORM LOADING   UNEQUAL LOADING

Figure 15.25 Load from narrow strip concentrated in center.

should be to the top left of the bearing cap; on the top-right roll, it should be at the top-right position.

The wrap on the bottom roll encompasses one-half of the roll circumference. As a result, the load vector is directly upward, or 90 degrees, to the passline. The best location for the primary radial-measurement point is in the vertical-downward position. The secondary radial measurement should be taken at 90 degrees to the primary. Since the strip tension is slightly forward (i.e., in the direction of strip movement), the secondary measurement should be taken on the recoiler-side of the bearing cap.

Because strip tension loads the bearings in the direction of the force vector, it also tends to dampen the vibration levels in the opposite direction, or 180 degrees, of the force vector. In effect, the strip acts like a rubber band. Tension inhibits movement and vibration in the direction opposite the force vector and amplifies the movement in the direction of the force vector. Therefore, the recommended measurement-point locations provide the best representation of the roll's dynamics.

In normal operation, the force or load induced by the strip is uniform across the roll's entire face or body. As a result, the vibration profile in both the operator- and drive-side bearings should be nearly identical.

Strip Width and Tracking
Strip width has a direct effect on roll loading and how the load is transmitted to the roll and its bearing support structures. Figure 15.25 illustrates a narrow strip that is tracking properly. Note that the load is concentrated on the center of the roll and is not uniform across the entire roll face. The concentration of strip tension or load in the center of the roll tends to bend the roll. The degree of deflection depends on the following: roll diameter, roll construction, and strip tension. Regardless of these three factors, however, the vibration profile is modified. Roll bending, or deflection,

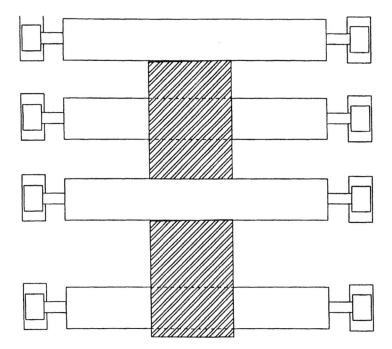

Figure 15.26 Roll loading.

increases the fundamental (1×) frequency component. The amount of increase is determined by the amount of deflection.

As long as the strip remains at the true centerline of the roll face, the vibration profile in both the operator- and drive-side bearing caps should remain nearly identical. The only exceptions are bearing rotational and defect frequencies. Figures 15.26 and 15.27 illustrate uneven loading and the resulting different vibration profiles of the operator- and drive-side bearing caps. This is an extremely important factor that can be used to evaluate many of the failure modes of continuous-process lines. For example, the vibration profile resulting from the transmission of strip tension to the roll and its bearings can be used to determine proper roll alignment, strip tracking, and proper strip tension.

Alignment

Process rolls must be properly aligned. The perception that they can be misaligned without causing poor quality, reduced capacity, and premature roll failure is incorrect. In the case of single rolls (e.g., bridle and furnace rolls), they must be perpendicular to the passline and have the same elevation on both the operator- and drive-side. Roll pairs such as scrubber/backup rolls must be absolutely parallel to each other.

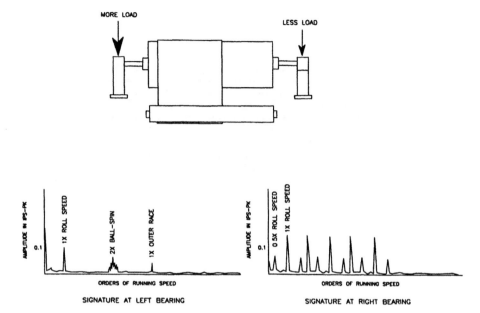

Figure 15.27 Typical vibration profile with uneven loading.

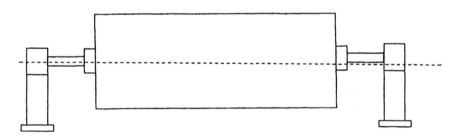

Figure 15.28 Vertically misaligned roll.

Single Rolls

With the exception of steering rolls, all single rolls in a continuous-process line must be perpendicular to the passline and have the same elevation on both the operator- and drive-side. Any horizontal or vertical misalignment influences the tracking of the strip and the vibration profile of the roll.

Figure 15.28 illustrates a roll that does not have the same elevation on both sides (i.e., vertical misalignment). With this type of misalignment, the strip has greater tension on the side of the roll with the higher elevation, which forces it to move toward the lower end. In effect, the roll becomes a steering roll, forcing the strip to one side of the centerline.

Scrubber Roll

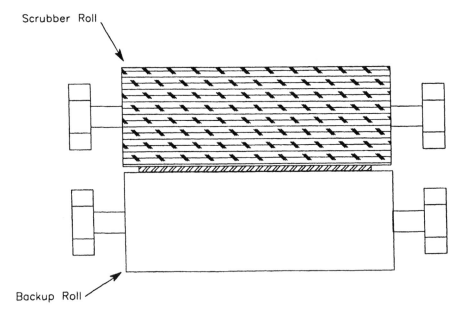

Backup Roll

Figure 15.29 Scrubber roll set.

The vibration profile of a vertically misaligned roll is not uniform. Because the strip tension is greater on the high side of the roll, the vibration profile on the high-side bearing has lower broadband energy. This is the result of damping caused by the strip tension. Dominant frequencies in this vibration profile are roll speed (1x) and outer-race defects. The low end of the roll has higher broadband vibration energy and dominant frequencies include roll speed (1x) and multiple harmonics (i.e., the same as mechanical looseness).

Paired Rolls

Rolls that are designed to work in pairs (e.g., Damming or Scrubber rolls) also must be perpendicular to the passline. In addition, they must be absolutely parallel to each other. Figure 15.29 illustrates a paired set of Scrubber rolls. The strip is captured between the two rolls and the counter-rotating brush roll cleans the strip surface.

Due to the designs of both the Damming and Scrubber roll sets, it is quite difficult to keep the rolls parallel. Most of these roll sets use a single pivot point to fix one end of the roll and a pneumatic cylinder to set the opposite end.

Other designs use two cylinders, one attached to each end of the roll. In these designs, the two cylinders are not mechanically linked and, therefore, the rolls do not maintain their parallel relationship. The result of nonparallel operation of these paired rolls is evident in roll life.

Scrubber Roll

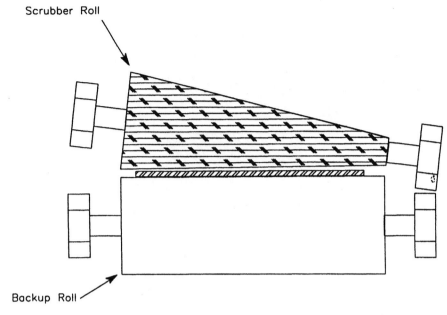

Backup Roll

Figure 15.30 Result of misalignment or nonparallel operation on brush rolls.

For example, the Scrubber/backup roll set should provide extended service life. However, in actual practice, the brush rolls have a service life of only a few weeks. After this short time in use, the brush rolls will have a conical shape, much like a bottle brush (see Figure 15.30). This wear pattern is visual confirmation that the brush roll and its mating rubber-coated backup roll are not parallel.

Vibration profiles can be used to determine if the roll pairs are parallel and, in this instance, the rules for parallel misalignment apply. If the rolls are misaligned, the vibration signatures exhibit a pronounced fundamental (1x) and second harmonic (2x) of roll speed.

Multiple Pairs of Rolls
Because the strip transmits the vibration profile associated with roll misalignment, it is difficult to isolate misalignment for a continuous-process line by evaluating one single or two paired rolls. The only way to isolate such misalignment is to analyze a series of rolls rather than individual (or a single pair of) rolls. This approach is consistent with good diagnostic practices and provides the means to isolate misaligned rolls and to verify strip tracking.

Strip Tracking
Figure 15.31 illustrates two sets of rolls in series. The bottom set of rolls is properly aligned and has good strip tracking. In this case, the vibration profiles acquired from

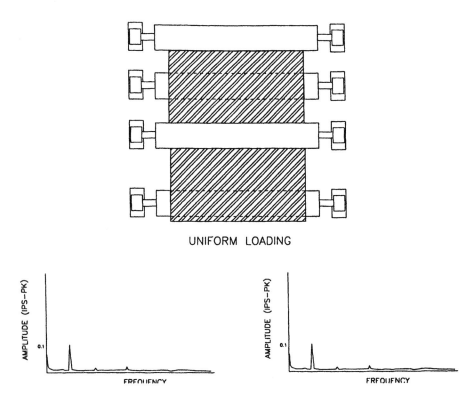

UNIFORM LOADING

Figure 15.31 Rolls in series.

the operator- and drive-side bearing caps are nearly identical. Unless there is a damaged bearing, all of the profiles contain low-level roll frequencies (1×) and bearing rotational frequencies.

The top roll set also is properly aligned, but the strip tracks to the bottom of the roll face. In this case, the vibration profile from all of the bottom bearing caps contains much lower level broadband energy and the top bearing caps have clear indications of mechanical looseness (i.e., multiple harmonics of rotating speed). The key to this type of analysis is the comparison of multiple rolls in the order in which they are connected by the strip. This requires comparison of both top and bottom rolls in the order of strip pass. With proper tracking, all bearing caps should be nearly identical. If the strip tracks to one side of the roll face, all bearing caps on that side of the line will have similar profiles. However, they will have radically different profiles compared to those on the opposite side.

Roll Misalignment

Roll misalignment can be detected and isolated using this same method. A misaligned roll in the series being evaluated causes a change in the strip track at the offending roll. The vibration profiles of rolls upstream of the misaligned roll will be identical on

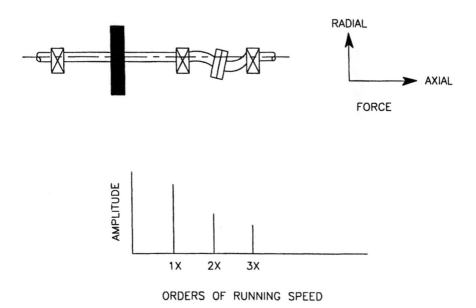

Figure 15.32 Bends that change shaft length generate axial thrust.

both the operator- and drive-side of the rolls. However, the profiles from the bearings of the misaligned roll will show a change. In most cases, they will show traditional misalignment (i.e., 1× and 2× components), but also will indicate a change in the uniform loading of the roll face. In other words, the overall or broadband vibration levels will be greater on one side than the other. The lower readings will be on the side with the higher strip tension and the higher readings will be on the side with less tension.

The rolls following the misalignment also show a change in vibration pattern. Since the misaligned roll acts as a steering roll, the loading patterns on the subsequent rolls show different vibration levels when the operator- and drive-sides are compared. If the strip track was normal prior to the misaligned roll, the subsequent rolls will indicate off-center tracking. In those cases where the strip was already tracking off-center, a misaligned roll either improves or amplifies the tracking problem. If the misaligned roll forces the strip toward the centerline, tracking improves and the vibration profiles are more uniform on both sides. If the misaligned roll forces the strip further off-center, the nonuniform vibration profiles will become even less uniform.

Shaft

A bent shaft creates an imbalance or a misaligned condition within a machine-train. Normally, this condition excites the fundamental (1×) and secondary (2×) running-speed components in the signature. However, it is difficult to determine the difference between a bent shaft, misalignment, and imbalance without a visual inspection.

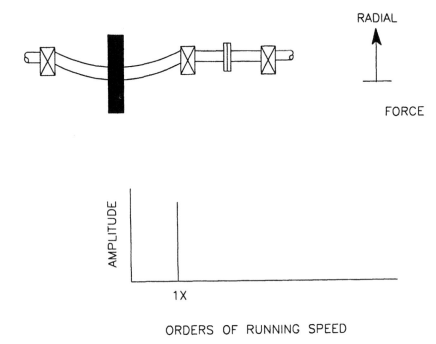

Figure 15.33 Bends that do not change shaft length generate radial forces only.

Figures 15.32 and 15.33 illustrate the normal types of bent shafts and the force profiles that result.

V-Belts

V-belt drives generate a series of dynamic forces, and vibrations result from these forces. Frequency components of such a drive can be attributed to sheaves and belts. The elastic nature of belts can either amplify or damp vibrations that are generated by the attached machine-train components.

Sheaves

Even new sheaves are not perfect and may be the source of abnormal forces and vibration. The primary sources of induced vibration due to sheaves are eccentricity, imbalance, misalignment, and wear.

Eccentricity

Vibration caused by sheave eccentricity manifests itself as changes in load and rotational speed. As an eccentric drive (Figure 15.34) sheave passes through its normal rotation, variations in the pitch diameter cause variations in the linear belt speed. An eccentric driven sheave causes variations in load to the drive. The rate at which such variations occur helps to determine which is eccentric. An eccentric sheave also may

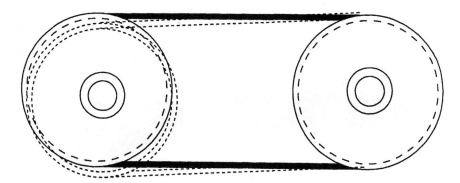

Figure 15.34 Eccentric sheaves.

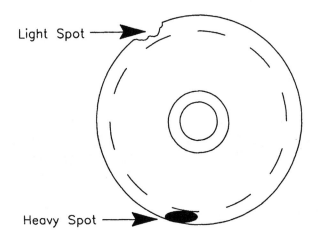

Figure 15.35 Light and heavy spots on an unbalanced sheave.

appear to be unbalanced. However, performing a balancing operation will not correct the eccentricity.

Imbalance
Sheave imbalance may be caused by several factors, one of which may be that it was never balanced to begin with. The easiest problem to detect is an actual imbalance of the sheave itself. A less obvious cause of imbalance is damage that has resulted in loss of sheave material. Imbalance due to material loss can be determined easily by visual inspection, either by removing the equipment from service or using a strobe light while the equipment is running. Figure 15.35 illustrates light and heavy spots that result in sheave imbalance.

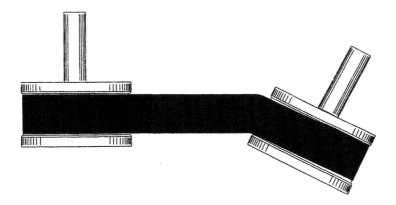

Figure 15.36 Angular sheave misalignment.

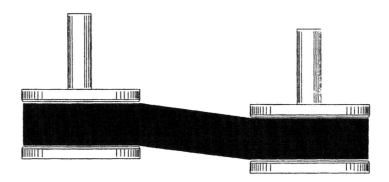

Figure 15.37 Parallel sheave misalignment.

Misalignment
Sheave misalignment most often produces axial vibration at the shaft rotational frequency (1×) and radial vibration at one and two times the shaft rotational frequency (1× and 2×). This vibration profile is similar to coupling misalignment. Figure 15.36 illustrates angular sheave misalignment and Figure 15.37 illustrates parallel misalignment.

Wear
Worn sheaves also may increase vibration at certain rotational frequencies. However, sheave wear is more often indicated by increased slippage and drive wear. Figure 15.38 illustrates both normal and worn sheave grooves.

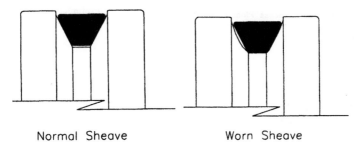

Normal Sheave   Worn Sheave

Figure 15.38 Normal and worn sheave grooves.

Belts

V-belt drives typically consist of multiple belts mated with sheaves to form a means of transmitting motive power. Individual belts, or an entire set of belts, can generate abnormal dynamic forces and vibration. The dominant sources of belt-induced vibrations are defects, imbalance, resonance, tension, and wear.

Defects
Belt defects appear in the vibration signature as subsynchronous peaks, often with harmonics. Figure 15.39 shows a typical spectral plot (i.e., vibration profile) for a defective belt.

Imbalance
An imbalanced belt produces vibration at its rotational frequency. If a belt's performance is initially acceptable and later develops an imbalance, the belt has most likely lost material and must be replaced. If imbalance occurs with a new belt, it is defective and also must be replaced. Figure 15.40 shows a spectral plot of shaft rotational and belt defect (i.e., imbalance) frequencies.

Resonance
Belt resonance occurs primarily when the natural frequency of some length of the belt is excited by a frequency generated by the drive. Occasionally, a sheave also may be excited by some drive frequency. Figure 15.41 shows a spectral plot of resonance excited by belt-defect frequency.

Belt resonance can be controlled by adjusting the span length, belt thickness, and belt tension. Altering any of these parameters changes the resonance characteristics. In most applications, it is not practical to alter the shaft rotational speeds, which also are possible sources of the excitation frequency.

Resonant belts are readily observable visually as excessive deflection, or belt whip. It can occur in any resonant mode so inflection points may or may not be observed

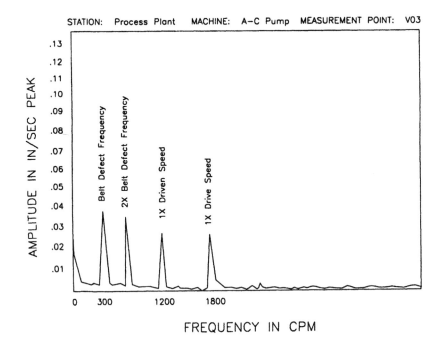

Figure 15.39 Typical spectral plot (i.e., vibration profile) of a defective belt.

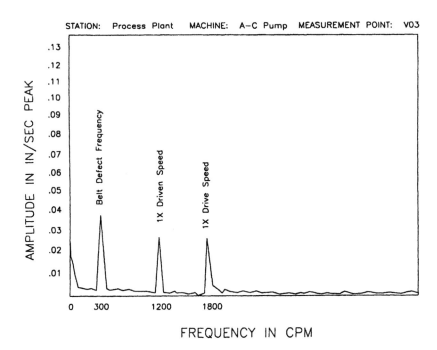

Figure 15.40 Spectral plot of shaft rotational and belt defect (i.e., imbalance) frequencies.

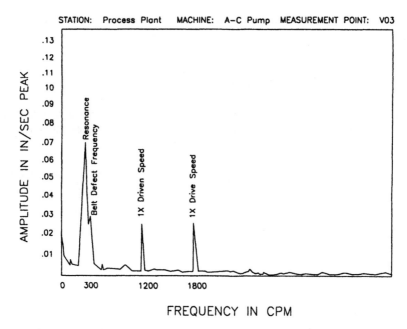

Figure 15.41 Spectral plot of resonance excited by belt-defect frequency.

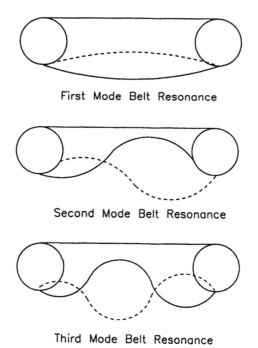

Figure 15.42 Examples of mode resonance in a belt span.

along the span. Figure 15.42 illustrates first-, second-, and third-mode resonance in a belt span.

Tension
Loose belts can increase the vibration of the drive, often in the axial plane.

In the case of multiple V-belt drives, mismatched belts also aggravate this condition. Improper sheave alignment can also compromise tension in multiple-belt drives.

Wear
Worn belts slip and the primary indication is speed change. If the speed of the driver increases and the speed of the driven unit decreases, then slippage is probably occurring. This condition may be accompanied by noise and smoke, causing belts to overheat and be glazed in appearance. It is important to replace worn belts.

Chapter 16

SIGNATURE ANALYSIS

Most failures of rotating and reciprocating machinery exhibit characteristic vibration profiles that are associated with specific failure modes. This phenomenon is due to the forcing function, caused by a developing defect, having a unique characteristic signature. None of the filtered bandwidth monitoring methods provides the means to detect and evaluate these unique profiles. Signature analysis provides this capability and its use is required in a comprehensive predictive maintenance program.

CHARACTERISTIC VIBRATION SIGNATURES

A vibration signature provides a clear, accurate snapshot of the unique frequency components generated by, or acting on, a machine-train. Such a signature is obtained by converting time-domain data into its unique frequency components using a fast Fourier transform (FFT). Such a vibration signature, referred to as frequency-domain data, is used in signature analysis to evaluate the dynamics of the machine.

Frequency-domain vibration signatures form the basis for any predictive maintenance program designed to detect, isolate, and verify incipient problems within a machine-train. These signatures are the basic tools used for in-depth analysis methods such as failure-mode, root-cause, and operating dynamics analyses. Operating dynamics analysis™, which is beyond the scope of this module, uses vibration data and other process parameters, such as flow rate, pressure, and temperature, to determine the actual operating condition of critical plant systems.

TYPES OF SIGNATURE ANALYSIS

In general, new or immature predictive maintenance programs are limited to comparative analysis or waterfall trending. Although these comparative techniques provide the ability to detect severe problems, they cannot be used to isolate and identify the forcing functions or failure modes. These methods also are limited in their ability to provide early detection of incipient problems.

As the predictive maintenance program matures, root-cause analysis and operating dynamics analysis™ methods can be used. With the addition of these more advanced diagnostic tools, vibration signatures become an even more valuable process performance improvement tool.

Automatic Trending Analysis

A predictive maintenance program utilizing a microprocessor-based vibration analyzer and a properly configured database automatically trends vibration data on each machine-train. In addition, it compares the data to established baselines and generates trend, time-to-failure, and alert/alarm status reports.

The use of just these standard capabilities greatly reduces unscheduled failures. However, these automated functions do not identify the root causes behind premature machine-train component failures. In most cases, more in-depth analysis allows the predictive analyst to identify the reason for pending failure and to recommend corrective actions to prevent a recurrence of the problem. Again, the specific microprocessor-based system used determines how much manual effort is required for more in-depth analysis.

More In-Depth Trending Analysis

More in-depth analysis is called for when the automatic trending analysis described in the previous section indicates that a machine-train is exhibiting excessive vibration. Obviously, machine-trains that are operating within acceptable boundaries do not require further investigation. Care should be taken, however, to ensure that the automated functions of the predictive maintenance system report abnormal growth trends as well as machine-trains that are actually in alarm.

Comparative Analysis (Waterfall Trending)

FFT signatures that are collected on a regular schedule provide a means of trending that can help the analyst identify changes in machine condition. Changes in the operating parameters, such as load, will directly affect the signatures generated by a machine.

Unlike trending analysis, which is based on broadband and narrowband data, comparative analysis is a visual comparison of the relative change of the machine-train's full vibration signature and its discrete frequency components over a period of time. Because vibration signatures are acquired at regular intervals in a predictive maintenance program, this form of trending is very effective in identifying changes in machine condition.

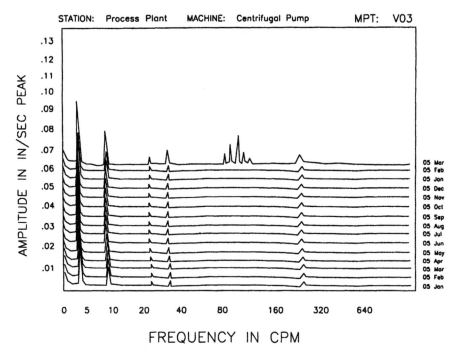

STATION: Process Plant   MACHINE: Centrifugal Pump   MPT: V03

Figure 16.1 Illustration of a waterfall plot.

Displaying the signatures in a waterfall or multiple-spectra display (sequentially by data-acquisition time) allows the analyst to easily see the relationship of each frequency component generated by the machine (see Figure 16.1). Any significant change in the amplitude of any discrete frequency is clearly evident in this type of display, which is used in many of the figures in subsequent sections.

Although comparative analysis can be used to help the analyst identify specific changes that are generated by process changes, each signature must be normalized for process variations. Therefore, as part of the acquired data set, the analyst must record the specific process conditions for each data set. With this information and the waterfall display of vibration signatures, the analyst can quantify the changes that result from variations of these parameters.

Developing problems within a machine-train can be identified by comparing the FFT signature to the following: (1) a baseline or reference signature, (2) previous signatures, or (3) industrial standards. This method determines if a potential problem exists and can be used to isolate within the machine-train the probable source of developing problems.

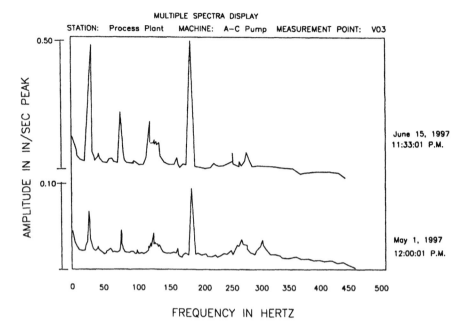

Figure 16.2 Comparison to baseline reference.

Baseline or Reference Signatures

A series of baseline or reference data sets should be taken for each machine-train included in a predictive maintenance program (Figure 16.2). These data sets are necessary to compare with trends, time traces, and FFT signatures that are collected over time. Therefore, baseline data sets must be representative of the normal operating condition of each machine-train in order to have value as a reference.

In integrated process plants where most machines are subject to variable operating conditions, this exercise requires more than one reference data set for each machine-train. To be of benefit, a series of baselines must be acquired from each machine-train, each of which should accurately represent a specific operating variable (i.e., product, machine setup, load, etc.). It is important that all data sets (whether baseline data or current operating data) be clearly identified in order to be useful. Current operations data must be compared to a reference data set having the same operating conditions (Figure 16.2).

Note that baseline references must be updated each time a machine-train is overhauled, replaced, or when a new process setup is established. A current set of valid reference data is essential when performing comparative analysis.

Table 16.1 Vibration Severity Standards*

Condition	Machine Classes (IPS-PK)			
	I	II	III	IV
Good operating condition	0.028	0.042	0.100	0.156
Alert limit	0.010	0.156	0.255	0.396
Alarm limit	0.156	0.396	0.396	0.622
Absolute-fault limit	0.260	0.400	0.620	1.000

* Applicable to a machine with running speed between 600 to 12,000 rpm. Narrowband setting: 0.3× to 3.0× running speed.

Machine Class Descriptions:

Class I   Small machine-trains or individual components integrally connected with the complete machine in its normal operating condition (i.e., drivers up to 20 hp).

Class II   Medium-sized machines (i.e., 20- to 100-hp drivers and 400-hp drivers on special foundations).

Class III   Large prime movers (i.e., drivers greater than 100 hp) mounted on heavy, rigid foundations.

Class IV   Large prime movers (i.e., drivers greater than 100 hp) mounted on relatively soft, lightweight structures.

Source: Derived by Integrated Systems, Inc., from ISO Standard 2372.

Nonbaseline Signatures

Visual comparison of two signatures can enable the analyst to determine if a problem is developing. As with the case of filtered energy data, all signatures must be normalized for process variables such as speed, load, etc., in order for comparisons to be valid. Direct comparison is useful only when both data sets reflect the same operating conditions or parameters.

Common-shaft analysis is used to identify the strongest vibration by visually comparing the signatures of all measurement points on a common shaft. It is a useful technique for isolating the source of abnormal vibrations. Although this method does not absolutely identify the problem, it reduces the number of machine components that must be inspected or evaluated to correct the problem.

Industrial Standards

One form of comparative analysis is direct comparison of the acquired data to industrial standards or reference values. The vibration severity standards presented in Table 16.1 were established by the International Standards Organization (ISO). These data are applicable for comparison with filtered narrowband data taken from machine-trains with true running speeds between 600 and 12,000 rpm. The values from the table include all vibration energy between a lower limit of 0.3× true running speed and an upper limit of 3.0×. For example, an 1800-rpm machine would have a filtered narrowband between 540 (1800 × 0.3) and 5400 rpm (1800 × 3.0). A 3600-rpm machine would have a filtered narrowband between 1080 (3600 × 0.3) and 10,800 rpm (3600 × 3.0).

Microprocessor Comparisons

Many of the microprocessor-based predictive maintenance systems also allow direct comparisons of the relative strengths of each frequency component. Such microprocessor comparisons do not require knowledge of the machine-train or vibration analysis techniques, but both data sets must be acquired under the same operating conditions. Increases in relative strength indicate more vibration and a developing problem in the machine-train.

Cross-machine comparison is an extremely beneficial tool to the novice analyst. Most vibration monitoring systems permit direct comparison of vibration data, both filtered window energy and complete signatures, acquired from two machines. This capability permits the analyst to directly compare a machine that is known to be in good operating condition with one that is perceived to have a problem. There are several ways that cross-machine comparisons can be made using microprocessor-based systems: multiple plots, ratio, and difference.

Multiple Plots
Two or more signatures can be shown on a single display. This method permits the analyst to directly compare the actual signatures generated at each measurement point on both the suspect and a reference machine-train. This multiple-signature display permits direct comparison of each frequency component within the signatures (Figure 16.3).

Ratio Analysis
With this technique, the signature from the suspect machine is divided by the signature of the reference machine, frequency by frequency. The resultant display shows the relative amplitude, both positive and negative, of each frequency component in the suspect machine-train (Figure 16.4). As an example, the display may indicate that the gear-mesh energy in the suspect machine is 40% higher than that in the reference machine (i.e., ratio = 1.4). With this information, the analyst can isolate specific machine components that are potential problems.

Difference Analysis
With the difference analysis technique, the signature of the reference machine is subtracted from that of the suspect machine, frequency by frequency. The resultant plot displays the difference value, positive and negative, of each frequency component within the two (see Figure 16.5).

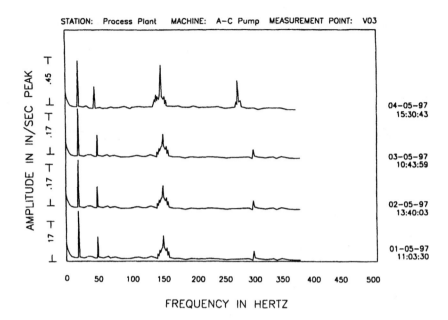

Figure 16.3 Multiple-signature display.

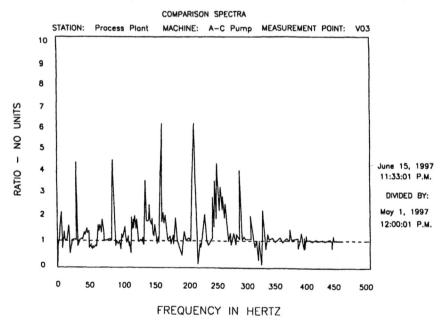

Figure 16.4 Ratio of two signatures.

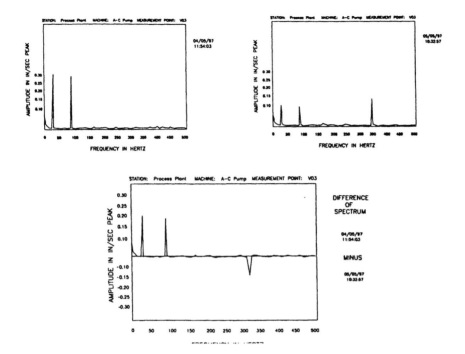

Figure 16.5 Difference of two signatures.

Chapter 17

ROOT-CAUSE ANALYSIS

Root-cause analysis provides the means to isolate and identify specific causes of abnormal vibration components observed for a machine-train. The trending and analysis methods discussed in the previous sections generally identify symptoms rather than causes of problems.

Traditional predictive maintenance programs do not typically incorporate root-cause analysis. Most programs are considered successful if they identify incipient problems in time to prevent severe damage or forced downtime. However, programs that do not include this type of analysis leave two questions unanswered: (1) Why did the problem occur? (2) Will it recur?

To achieve maximum benefit from a predictive maintenance program, it is not enough to predict failure in time to repair it without affecting production. In most instances, problems can be solved at minimal expense through the use of root-cause analysis. However, problems are often ignored for quite some time and machine failures allowed to recur. The net result is that machine life can be drastically shortened, maintenance cost is increased, and available production time is reduced. In short, ignoring chronic problems, no matter how small they seem, costs a typical plant a substantial amount of actual, or potential, revenue every year.

DETERMINING WHY PROBLEMS OCCUR

Machine-train failures do not occur without a reason. To solve the problem, the reason for failure must be found and corrected. Root-cause analysis is based on machine-train operation and how its dynamics affect the vibration spectrum. However, the analyst may have to evaluate the entire process system to determine the reason behind a chronic machine-train failure. If the chronic problem is system related rather than machine-train related, the knowledge of process dynamics required to perform the analysis may be beyond the capabilities of the predictive maintenance staff. In this case, the assistance of the plant engineering staff is required.

In extreme cases, the experience of the plant engineering staff will not be sufficient to cost effectively resolve a chronic problem. Most plant engineers are generalists within a specific discipline (e.g., mechanical, electrical). As such, they are not, and should not be expected to be, specialists in machinery or process design and application.

Many of the chronic machinery problems that occur daily require special knowledge of the specific machinery and system to find the solution. In most cases, the first line of support for this expertise is the original equipment manufacturer. The original manufacturer should certainly have the required machine design knowledge. The manufacturer may, however, be biased. In many cases, an unbiased consultant can provide a cost-effective solution to a chronic problem.

WHEN TO PERFORM THE ANALYSIS

True root-cause analysis is not necessary for many machine-train problems. However, if the problem is chronic and recurs more often than expected, such an analysis is essential.

In many cases, recurring machine-train problems can be traced to a system or process rather than something mechanical. However, chronic mechanical problems do exist and usually can be traced to improper installation, application, or maintenance.

A classic example of a process-related problem is a recurring centrifugal air compressor problem experienced by one of our clients. In this instance, three 4-stage centrifugal compressors, each rated at 2150 cfm at 100 psig on a common header, exhibited chronic failure problems. In a period of approximately 12 months, all three compressors required major rebuilds and several minor repairs.

The predictive maintenance program identified each failure and a failure-mode analysis isolated and identified the specific failure point. Nevertheless, the problem continued to recur. A root-cause analysis indicated that the real problem was not mechanical (i.e., compressor/motor), but was a system (i.e., plant air) problem. A redesign of the plant air system eliminated the problem and normal machine life was restored.

METHODOLOGY

The methodology used in root-cause analysis is basically the same as that used for failure-mode analysis. The major difference is that for root-cause analysis the vibration signature is only one part of the data required to isolate the problem. In addition to vibration data, complete process data (e.g., flow, pressure, temperature), detailed process system-design data, and machine-design data are required. Sometimes, testing utilizing techniques other than vibration analysis is required. However, the concept of root-cause analysis is to understand the relationship of all components within the system so that the cause of a problem can be identified.

The premise of root-cause analysis, like failure-mode analysis, is that there is an identifiable reason or reasons for each abnormal occurrence in machine-train or system operation. However, each application is unique and the methods used vary with the specific process system or application. Therefore, root-cause analysis requires a complete knowledge of machine and process system design, operation, and typical failure modes. While many chronic problems are unique to a particular plant, many are commonly found in other applications, plants, and/or industries.

Unfortunately, obtaining a good grasp of root-cause analysis techniques requires much training and experience. In addition to formal technical education and plant experience, a good experience base in both normal and abnormal systems operation is required to cost effectively and accurately identify and correct most typical chronic problems.

Because of this, it is generally more cost-effective to use a consultant to resolve many chronic problems rather than training in-house staff. However, simple root-cause analysis can and should be included in most predictive maintenance programs. Seminars that provide the basic knowledge required to perform simple analyses are available from a number of consultants and predictive maintenance system manufacturers. Simple root-cause analysis is certainly within the capabilities of most predictive maintenance personnel and, with some technical support from outside consultants, most chronic problems can be resolved.

SUSCEPTIBILITY TO CHRONIC PROBLEMS

Some machinery and process systems are more susceptible to chronic process- or application-induced problems than are others. Compressors, pumps, and fans are examples of machinery that commonly exhibit chronic problems as a direct result of process or application problems. These machines are extremely sensitive to process variables.

Chronic bearing failure is probably the most common machine/system-induced problem. By design, bearings are the weakest link in most machinery. A high failure rate is a definite indication of a system/machine problem and chronic failures should not be ignored because they are definite indicators that machine-train life is being adversely affected. If bearings fail, other mechanical damage also may be present.

DIAGNOSTIC TECHNIQUES

The following diagnostic and data techniques are discussed in this section: common-shaft analysis, shaft deflection, and data normalization.

Common-Shaft Analysis

Vibration analysis requires the use of proper evaluation techniques that permit the analyst to understand fully the operating condition of each machine-train and its associated system. However, the natural tendency is to evaluate each measurement point individually or to limit evaluation to a single machine component, such as an electric motor. While this approach provides some insight, it cannot provide a complete picture of the machine-train's dynamics.

A fundamental requirement of good analysis technique is to evaluate each machine-train using a common-shaft approach rather than using data from individual measurement points or machine-train components not found on a common shaft. With this approach, all data acquired from the entire machine-train—from the first outboard motor bearing to the outboard bearing of the final driven component—are evaluated as a series of common shafts.

What Is a Common Shaft?

The term *common shaft* refers to each individual shaft that exists in a machine-train. For example, there are two common shafts in a machine-train consisting of an electric motor that is directly coupled to a single-reduction gearbox that is directly coupled to a fan or other machine component. The first extends from the outboard motor bearing to the outboard bearing of the high-speed (input) gearbox shaft. The second common shaft extends from the outboard bearing of the low-speed (output) gearbox shaft to the outboard bearing of the fan or driven machine.

Figure 17.1 illustrates a double-reduction gearbox drive-train. In this configuration, there are three common shafts. The first begins at the outboard motor bearing and ends at the outboard bearing of the gearbox input shaft. The second begins at the inboard bearing of the intermediate gear shaft and ends with the intermediate gear shaft outboard bearing. The final common shaft begins with the inboard (left) bearing of the gearbox output shaft and continues through the pump impeller.

Data Organization

Data obtained from a common-shaft analysis should be organized so that the deflection and dynamics of all shafts within the extended machine-train can be evaluated as a single unit. For example, the primary radial data points from all bearing caps on a common shaft are evaluated as one unit; the secondary data points as a second unit; and the axial data points as the final unit. After each of the individual common shafts has been evaluated, the analyst can compare the dynamics of each to gain a clear understanding of the overall dynamics of the entire machine-train.

Advantages

The use of common-shaft analysis greatly enhances an analyst's ability to detect and diagnose machine-train problems. The advantages of this approach include the ability

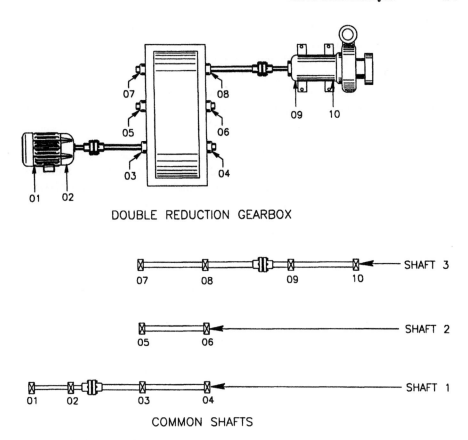

Figure 17.1 Example of common shafts in machine-trains.

to confirm common failure modes; obtain a clear picture of the shaft displacement, both internal and external; and isolate any abnormal forcing functions that may exist.

Comparing the amplitude of individual frequency components at the various measurement points along a common shaft is a simple means of isolating the location of developing machine-train problems. The energy of each vibration component is greatest at the machine component or source of problem. This is because the machine-train housing absorbs the energy as the data-measurement location moves away from the source.

Therefore, comparing the strength of each frequency component at various points along the common shaft is a means of isolating the source. Since most of the unique frequency components have a direct relationship to the running speed of a specific shaft, monitoring all measurement points on a common shaft simplifies comparative analysis of the unique frequency components created by that shaft.

Even though this technique does not identify the specific failure mode or problem (e.g., misalignment, imbalance), it does isolate the source or location of the problem. With this information, specific problems are much easier to isolate and identify. For example, a comparison of the 1×, 2×, and 3× components at each measurement point on a common shaft can be used to locate potential misalignment problems. The amplitude of each abnormal component is greatest at the source of vibration and decreases as the distance from the source increases.

Common-shaft analysis provides the ability to confirm or eliminate potential failure modes. For example, an electric motor may exhibit an abnormal second-harmonic (2×) frequency that is normally associated with parallel misalignment. If analysis is limited to individual machine components, in this case the motor, there is no way to confirm misalignment between the motor and its driven unit. When common-shaft techniques are used, data from the driven unit is used to confirm or eliminate mis- alignment. In this instance, the driven unit must also contain abnormal 2× vibration to confirm misalignment between the motor and driven unit.

Machines in Series

The common-shaft approach can be expanded to include analysis of machines in series, for example, the furnace roll of a continuous-process line or mill stands in a hot strip or tandem mill. After completing an analysis of the individual machine-trains in this example, the analyst should compare the dynamics of each machine-train to those connected to them (Figure 17.2). In many cases, the only point of connection is the strip, but this is more than enough to provide a direct mechanical connection and to interlace the dynamics of the machines in series.

This approach is essential on all continuous-process lines and where two or more machines are installed in series. By analyzing the entire system as a unit, the analyst can better understand the dynamics of each machine. The use of this approach can be illustrated with the misalignment that is frequently observed in single-reduction gear- boxes that drive rolls found in the drive-train of a continuous-process line. Often, roll misalignment or improper strip tracking or tension is the real source of the problem that is diagnosed as misalignment in the gear set.

The abnormal loading of the rolls that occurs as a result of roll misalignment or improper strip tracking is transmitted into the gearbox and, in effect, misaligns the gears. By evaluating the interaction of these machines in series, the analyst can clearly see that the source of the misalignment is outside of the gearbox and can track the problem to its source.

Machines in Parallel

In applications where two or more machines are installed and operated in parallel, a similar common-shaft approach should be used. After analyzing each of the machine- trains, the analyst should compare the results to determine the interaction of the com-

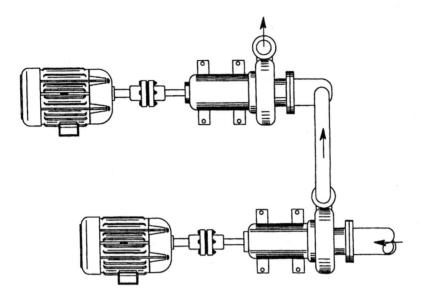

FLOW ←

Figure 17.2 Machines in series should be analyzed as a unit.

plete system (Figure 17.3). In many cases, systems that use multiple machines in parallel are subject to moderate to severe process instability. This instability is created by improper piping configurations, improper isolation of nonrunning or standby machines, inlet starvation, and a variety of other causes. The only means of isolating these problems is to analyze the entire system, not just the individual machine-trains.

The probability of two or more machines generating the exact same vibration profile is extremely unlikely, but many system problems create a common profile in the machines that make up the system. By comparing the profiles of all machines in parallel, the analyst can quickly see that the problem is system related rather than machine related.

Shaft Deflection

Analysis of shaft deflection is a fundamental diagnostic tool. If the analyst can establish the specific direction and approximate severity of shaft displacement, it is much easier to isolate the forcing function. For example, when the discharge valve on an end-suction centrifugal pump is restricted, the pump's shaft is displaced in a direction opposite to the discharge volute. Such deflection is caused by the back-pressure generated by the partially closed valve. Most of the failure modes and abnormal operating dynamics that affect machine reliability force the shaft from its true centerline. By

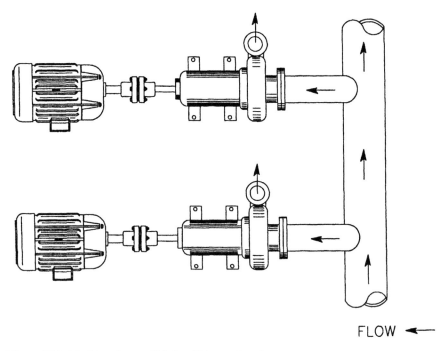

FLOW ←

Figure 17.3 Machines in parallel should be analyzed as a group.

using common-shaft diagnostics, the analyst can detect deviations from normal operating condition and isolate the probable forcing function.

Data Normalization

Data acquired as part of a regular vibration-monitoring program must be normalized before an effective analysis can be performed. It is virtually impossible to properly evaluate the machine condition or to detect abnormal behavior without normalizing the data.

Normalization is required to eliminate the effects of process changes in the vibration profiles. At a minimum, each data set must be normalized for speed, load, and the other standard process variables. Normalization allows the use of trending techniques, or the comparison of a series of profiles generated over time.

Regardless of the machine's operating conditions, the frequency components should occur at the same location when comparing normalized data for a machine. Normalization allows the location of frequency components to be expressed as an integer . multiple of shaft running speed, although fractions sometimes result. For example, gear-mesh frequency locations are generally integer multiples (5×, 10×, etc.) and bearing-frequency locations are generally noninteger multiples (0.5×, 1.5×, etc.). Plotting the vibration signature in multiples of running speed quickly differentiates the unique frequencies that are generated by bearings from those generated by gears, blades, and other components that are integers of running speed.

Speed

When normalizing data for speed, all machines should be considered to be variable speed—even those classified as constant speed. Speed changes due to load occur even with simple "constant-speed" machine-trains, such as electric-motor-driven centrifugal pumps. Generally, the change is relatively minor (between 5 and 15%), but it is enough to affect diagnostic accuracy. This variation in speed is enough to distort vibration signatures, which can lead to improper diagnosis.

With constant-speed machines, an analyst's normal tendency is to normalize speed to the default speed used in the database setup. However, this practice can introduce enough error to distort the results of the analysis because the default speed is usually an average value from the manufacturer.

For example, a motor may have been assigned a speed of 1780 rpm during setup. The analyst then assumes that all data sets were acquired at this speed. In actual practice, however, the motor's speed could vary the full range between locked rotor speed (i.e., maximum load) to synchronous (i.e., no-load) speed. In this example, the range could be between 1750 and 1800 rpm, a difference of 50 rpm. This variation is enough to distort data normalized to 1780 rpm. Therefore, it is necessary to normalize each data set to the actual operating speed that occurs during data acquisition rather than using the default speed from the database.

Take care when using the vibration analysis software provided with most microprocessor-based systems to determine the machine speed to use for data normalization. In particular, do not obtain the machine speed value from a display screen plot (i.e., on-screen or print-screen) generated by a microprocessor-based vibration analysis software program. This value is not to be used because the cursor position does not represent the true frequency of displayed peaks, but is instead an average value (see Figure 17.4). The graphics packages in most of the programs use an average of four or five data points to plot each visible peak. This technique is acceptable for most data-analysis purposes, but can skew the results if used to normalize the data. The approximate machine speed obtained from such a plot is usually within 10% of the actual value, which is not accurate enough to be used for speed normalization. Instead, use the peak search algorithm and print out the actual peaks and associated speeds.

Load

Data also must be normalized for variations in load. Where speed variations result in a right or left shift of the frequency components, variations in load change the amplitude. For example, the vibration amplitude of a centrifugal compressor taken at 100%

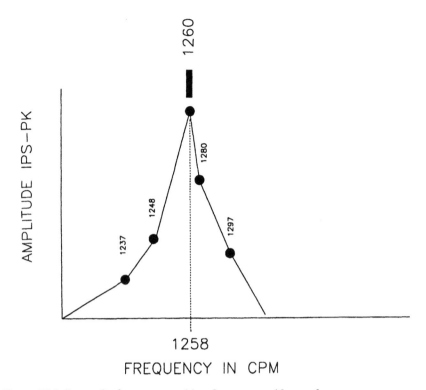

Figure 17.4 Screen display cursor position does not provide true frequency.

load is substantially lower than the vibration amplitude in the same compressor operating at 50% load.

In addition, the effect of load variation is not linear. In other words, the change in overall vibration energy does not change by 50% with a corresponding 50% load variation. Instead, it tends to follow more of a quadratic relationship. A 50% load variation can create a 200%, or a factor of 4, change in vibration energy.

Note that none of the comparative trending or analysis techniques can be used on variable-load machine-trains without first normalizing the data. Again, since even machines classified as constant-load machines operate in a variable-load condition, it is good practice to normalize all data to compensate for load variations utilizing the proper relationship for the application.

Other Process Variables

Other variations in a process or system have a direct effect on the operating dynamics and vibration profile of the machinery. In addition to changes in speed and load, other process variables affect the stability of the rotating elements, induce abnormal distribution of loads, and cause a variety of other abnormalities that directly impact diagnostics. Therefore, each acquired data set should include a full description of the machine-train and process system parameters.

As an example, abnormal strip tension or traction in a continuous-process line changes the load distribution on the process rolls that transport a strip through the line. This abnormal loading induces a form of misalignment that is visible in the roll and its drive-train's vibration profile.

Part III

RESONANCE AND CRITICAL SPEED
ANALYSIS

Chapter 18

INTRODUCTION

Resonance is a large-amplitude vibration caused by a small periodic stimulus having the same, or nearly the same, period as the system's natural vibration. In other words, a low-level energy source may excite a natural frequency and cause a substantial increase in its amplitude. Every machine-train has at least one natural vibration frequency.

Resonance is a very destructive vibration and, in most cases, will cause major damage to the machine or support structure. An example of resonance is a tuning fork. If you activate a tuning fork by striking it sharply, the fork vibrates rapidly. As long as it is held suspended, the vibration decays with time. However, if you place it on a desk top, the fork may excite the natural frequency of the desk, which dramatically amplifies the vibration energy.

The same thing can occur if one or more of the running speeds of a machine excites the natural frequency of the machine or its support structure. The key point to remember is that a very low amplitude energy source can cause a large vibration amplitude when its frequency coincides with the natural frequency of a machine or structure. Higher levels of input energy can cause catastrophic, near-instantaneous failure of the machine or structure.

Chapter 19

TYPES OF RESONANCE

Two major classifications of resonance are found in most manufacturing and process plants: static and dynamic. Both types exhibit a broad-based, high-amplitude frequency component when viewed in a fast Fourier transform (FFT), or frequency-domain, vibration signature. Unlike meshing or passing frequencies, the resonance frequency component does not have modulations or sidebands. Instead, resonance is displayed as a single, clearly defined peak, which represents a large quantity of energy. Such a peak is illustrated in Figure 19.1.

The high-energy content results from the amplitude of the peak as well as the broad area under the peak. This combination of high peak amplitude and area is typical of most resonance problems. The damping system associated with a resonance frequency is indicated by the sharpness or width of the response curve, ω_n, when measured at the half-power point. R_{MAX} is the maximum resonance and $R_{MAX}/\sqrt{2}$ is the half-power point for a typical resonance response curve.

To determine system damping, we must determine the maximum response. This is the response at the resonant frequency as indicated by the maximum value of $R_{velocity}$ or $(R_v)_{MAX}$ (a dimensionless velocity response factor), which is defined by Q as shown in the following equation:

$$Q = (R_v)_{MAX} = \frac{1}{2}\varsigma$$

where ς is fraction of critical damping. The maximum dimensionless acceleration factor, $(R_a)_{MAX}$, and dimensionless displacement factor, $(R_d)_{MAX}$, responses are slightly larger and can be calculated as:

$$(R_d)_{MAX} = (R_a)_{MAX} = \frac{(R_v)_{MAX}}{\left(1 - \varsigma^2\right)^{1/2}}$$

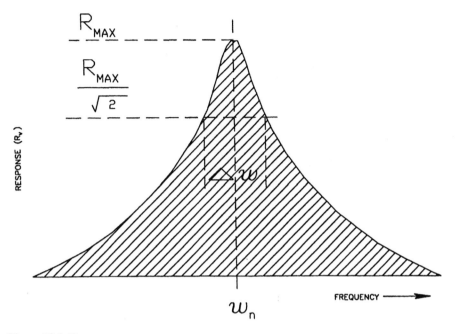

Figure 19.1 Resonance response.

The damping in a system is indicated by the sharpness or width of the response curve in the vicinity of a resonant frequency ω_n. The width is designated as a frequency increment, $\Delta\omega$, measured at the half-power point where the value of R is equal to $\dfrac{R_{MAX}}{\sqrt{2}}$. This also is illustrated in Figure 19.1. Where the values of ζ are less than 0.1, the damping of the system can be approximated by:

$$\frac{\Delta\omega}{\omega_n} = \frac{1}{Q} 2\zeta$$

STATIC RESONANCE

Static resonance is a function of the natural frequency of nondynamic, or stationary, machine components (e.g., casings and bearing support pedestals) and structural members (i.e., beams, piping, etc.). When one or more of the natural frequencies of a stationary structure is energized or excited, it resonates.

Because static resonance is a nondynamic phenomenon, it is generally not associated with the primary running speed of any associated machinery. Rather, the source of static resonance can be any energy source that coincides with the natural frequency of any stationary component. For example, an I-beam support on a continuous annealing

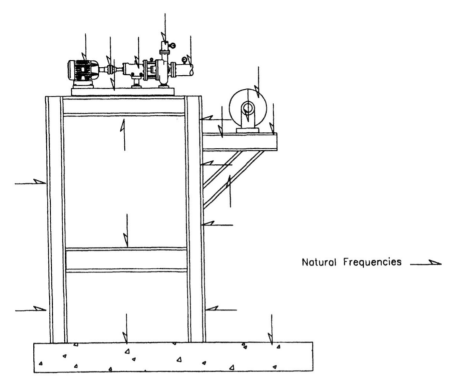

Figure 19.2 Typical discrete natural frequency locations in structural members.

line may be energized by the running speed of a roll. However, it also can be made to resonate by a bearing frequency, overhead crane, or other such energy source.

The resonant frequency depends on the mass, stiffness, and span of the excited member. In general terms, the natural frequency of a structural member is inversely proportional to its mass and stiffness. In other words, a large turbocompressor's casing will have a lower natural frequency than that of a small end-suction centrifugal pump.

Figure 19.2 illustrates a typical structural support system and the discrete natural frequency locations. Each of the arrows indicates a structural member or stationary machine component having a unique natural frequency. Note that each time a structural span is broken or attached to another structure, the stiffness changes. As a result, the natural frequency of that segment also changes.

While most stationary machine components move during normal operation, they are not always resonant. Some degree of flexing or movement is common in stationary machine-trains and structural members. The amount of movement depends on the spring constant, or stiffness, of the member.

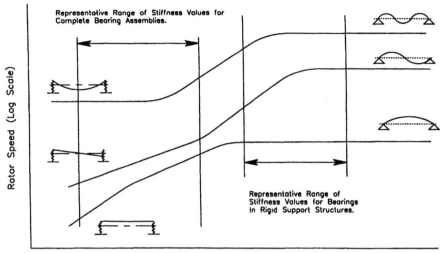

Figure 19.3 Rotor support stiffness versus critical rotor speed.

DYNAMIC RESONANCE

When the natural frequency of a rotating (i.e., dynamic) structure, such as a bearing or a rotor assembly in a fan, is energized, the rotating machine element resonates. This phenomenon is called dynamic resonance and the rotor speed at which it occurs is the critical speed.

Figure 19.3 illustrates a typical critical speed, or dynamic resonance, plot. The graph shows the relationship between rotor-support stiffness (X-axis) and rotor speed (Y-axis). Rotor-support stiffness depends on the geometry of the rotating element (i.e., shaft and rotor) and the bearing-support structure. These are the two dominant factors that determine the response characteristics of the rotor assembly.

In most cases, running speed is the forcing function that excites the natural frequency of the dynamic component. As a result, rotating equipment is designed to operate at primary rotor speeds that do not coincide with the rotor assembly's natural frequencies. As with static components, dynamic machine components have one or more natural frequencies that can be excited by an energy source that coincides with, or is in proximity to, that frequency.

High amplitudes of the rotor's natural frequency are strictly speed dependent. If the frequency of the energy source, in this case speed, changes to a value outside the resonant zone, the abnormal vibration disappears.

As with static resonance, the actual natural frequencies of dynamic members depend on the mass, bearing span, shaft and bearing-support stiffness, freedom of movement, and other factors that define the response characteristics of the rotor assembly (i.e., rotor dynamics) under various operating conditions.

In most cases, dynamic resonance appears at the fundamental running speed or one of the harmonics of the excited rotating element. However, it also can occur at other frequencies. For example, a rotor assembly with a natural frequency of 1800 rpm cannot operate at speeds between 1980 and 1620 rpm (± 10%) without the possibility of exciting the rotor's natural frequency.

Most low- to moderate-speed machinery is designed to operate below the first critical speed of the rotor assembly. Higher speed machines may be designed to operate between the first and second, or second and third, critical speeds of the rotor assembly. As these machines accelerate through the resonant zones or critical speeds, their natural frequency is momentarily excited. As long as the ramp rate limits the duration of excitation, this mode of operation is acceptable. However, care must be taken to ensure that the transition time through the resonant zone is as short as possible.

Note that critical speed should not be confused with the mode shape of a rotating shaft. Deflection of the shaft from its true centerline (i.e., mode shape) elevates the vibration amplitude and generates dominant vibration frequencies at the rotor's fundamental and harmonics of the running speed.

However, the amplitude of these frequency components tends to be much lower than those caused by operating at a critical speed of the rotor assembly. Also, the excessive vibration amplitude generated by operating at a critical speed disappears when the speed is changed, but those caused by mode shape tend to remain through a much wider speed range or may even be independent of speed.

Confirmation Analysis

In most cases, the occurrence of dynamic resonance can be quickly confirmed. When monitoring phase and amplitude, resonance is indicated by a 180-degree phase shift as the rotor passes through the resonant zone. Figure 19.4 illustrates a dynamic resonance at 500 rpm, which shows a dramatic amplitude increase in the frequency-domain display. Resonance is confirmed by the 180-degree phase shift in the time-domain plot. Note that the peak at 1200 rpm is not resonance. The absence of a phase shift, coupled with the apparent modulations in the FFT, eliminates the possibility that this peak is resonance related.

Common Confusions

Vibration analysts often confuse resonance with other failure modes. Many of the common failure modes tend to create abnormally high vibration levels that appear to

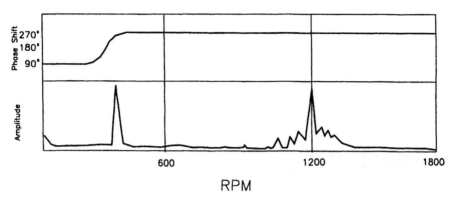

Figure 19.4 Dynamic resonance phase shift.

be related to a speed change. Therefore, analysts tend to miss the root cause of these problems.

Dynamic resonance generates abnormal vibration profiles that tend to coincide with the fundamental (1x) running speed or one or more of the harmonics. This often leads the analyst to incorrectly diagnose the problem as imbalance or misalignment.

Chapter 20

EXAMPLES OF RESONANCE

STATIC RESONANCE

Many machine-trains and production systems are subject to static and dynamic resonance. This section discusses some specific examples for each type.

Examples of machinery that exhibit static resonance are variable-speed machines, continuous-process lines, and deck-mounted machine trains.

Variable-Speed Machines

A variety of variable-speed process machinery, such as four-high rolling mills used by the steel industry (Figure 20.1) are operated over a wide range of running speeds. Because their normal mode of operation tends to excite one or more of the machine's natural frequencies, these machines are prime examples of equipment that experiences static resonance.

Waterfall data, such as that taken from a typical cold reduction mill, clearly displays the transitions through the resonance zones of a four-high mill. These zones occur as the mill accelerates from dead-stop, and decelerates from full speed to dead-stop. Some of the resonance zones are caused by excitation of the natural frequencies of the mill stand or other stationary members of the mill. Others are the result of dynamic resonance created by the excitation of the natural frequency of a roll or other rotating member within the mill.

Without a clear understanding of the specific natural frequencies of a process system, it is difficult to separate the static and dynamic resonance exhibited by a waterfall plot such as the one shown in Figure 20.2. This figure is a typical waterfall plot of a complete production cycle for a cold reduction mill. Note how the running speed of the

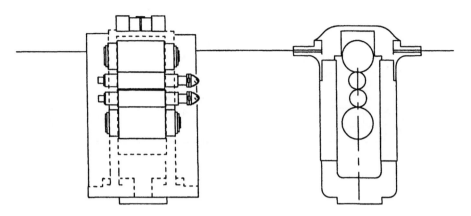

Figure 20.1 Variable-speed four-high rolling mill.

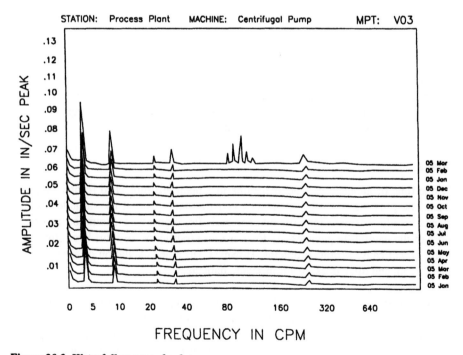

Figure 20.2 Waterfall or cascade plot.

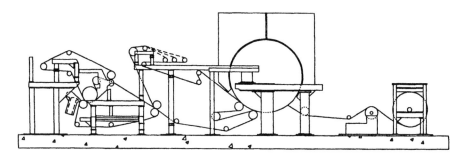

Figure 20.3 Continuous-process line.

rolls, gears, and other mill components passes through a number of resonant zones as the mill accelerates. These resonance zones, displayed as broad-based peaks, are clearly visible as mill speed increases from left to right of the horizontal axis.

Continuous-Process Lines

All continuous-process lines, such as plating lines, paper machines, etc., are subject to resonance due to the excitation of one or more natural frequencies of their support structure. Figure 20.3 illustrates a continuous-process line.

Deck-Mounted Machine-Trains

Any machine-train that is mounted on a deck-plate rather than a solid concrete foundation is subject to resonance problems. If the stiffness of the deck-plate is inadequate or the support span is too great, the normal result is static resonance created by the excitation of the deck-plate's natural frequency.

Others

At one time or another, all machine-trains in a plant are subject to resonance problems. The proximity of other machines and transients caused by variable running speeds greatly increases the potential for periodic, momentary excitation of one or more of the natural frequencies. As long as these transients are short lived, they normally do not cause serious problems. However, sustained excitation of a natural frequency can, and often does, result in severe damage.

In most cases, resonance is limited to the casing or support structure of the machine-train. The resulting vibration typically has a low frequency and may exhibit extremely high amplitudes. Gearboxes, compressors, pumps, and other machine types are particularly susceptible to this form of resonance. Because the source of excitation is often external to the monitored machine, static resonance is generally difficult to isolate.

DYNAMIC RESONANCE

Most of the machine-trains used in a plant are susceptible to dynamic resonance. It is especially prevalent in variable-speed machine-trains that are operated over a wide range of speeds. However, even constant-speed machines, such as fans and blowers, are prime candidates for resonance problems. Rolling mills, which are variable-speed machines, also are prime candidates for dynamic resonance.

Fans and Blowers

Dynamic resonance is one of the most common failure modes of fans and blowers. While most fans are operated at or near constant speed, it is possible to create situations where the speed of rotation coincides with the rotor's natural frequency. Although all fans and blowers are susceptible, cantilevered or overhung designs are the most likely candidates for resonance or critical speed problems.

Typical fans and blowers are designed to operate at speeds 10 to 15% below the rotor's first critical speed. As long as the fan's speed and the rotor's mass remain constant, this design practice does not create a problem. However, when either speed or mass changes, serious problems may result.

Many fans are belt driven. As a result, the sheave ratio may be changed to increase speed. In some cases, this change in ratio and, hence, speed is unintentional. For example, a millwright might replace a damaged sheave with one of different diameter. In other cases, the speed may be raised in an attempt to increase flow or pressure. In either case, the result is the same. The new fan speed may coincide with the first critical speed of the rotor assembly and severe, potentially destructive vibration may occur.

Another common problem associated with fans and blowers is an increase in rotor mass. In the dirty plant environment, the rotor assemblies in fans and blowers tend to accumulate dirt, moisture, and other contaminants. This phenomenon, called plate-out, increases the mass of the rotating element. Because the natural frequency of the rotor is dependent on its mass, this increase changes the natural frequency. As the mass increases, the natural frequency becomes lower. If the mass changes enough, the first critical of the rotor assembly may coincide with the design running speed. The result is an increase in vibration amplitude at running speed.

Rolling Mills

As mentioned in the variable-speed machine discussion all hot and cold reduction rolling mills are highly susceptible to dynamic resonance. Each of the rolls has a natural frequency determined by its installed configuration. The natural frequency of each roll depends on a number of variables that change during normal operation of the mill.

Variables, such as roll bending, roll force, and balancing force, will change the natural frequency of each roll. As a result, it is extremely difficult to isolate the specific roll that is being affected by resonance. Many of the chatter problems associated with cold reduction and temper mills are caused by dynamic resonance. Chatter is caused by gauge deviation in the strip. Third and fifth octave chatter problems are, in many cases, the excitation of the natural frequencies of the work and backup rolls (dynamic resonance) or mill stand (static resonance).

Chapter 21

TESTING FOR RESONANCE

The purpose of resonance testing is to isolate the machine component that is being excited and to determine the source of the excitation force.

STATIC RESONANCE

Static resonance testing is limited to structural members or machine components that do not have dynamic physical properties (i.e., properties that change with speed or time). Such structures include piping, machine casings, machine supports, deck-plates, and other structural members.

During testing, the natural frequencies of the entire system are compared with the vibration, or forcing, frequencies on an interference (i.e., Campbell) diagram to determine if the system is resonant. Figure 21.1 illustrates such a diagram.

In most cases, evidence of a potential static resonance problem will be found in the routine frequency- and time-domain vibration data that are collected as part of a predictive maintenance program. These data will contain high-amplitude, high-energy frequency components that cannot be explained or identified as a specific dynamic force generated by the machine-train or its systems. The component generated by potential static resonance may be at any frequency from 1 Hz to 30 kHz, but will rarely fall at the fundamental (1×) or any harmonic of running speed.

Isolating the Natural Frequency

In most cases, identifying the specific structural member or static machine component being excited is very difficult. In a typical structure, there are a large number of natural frequencies with each corresponding to a specific structural member or span. As a

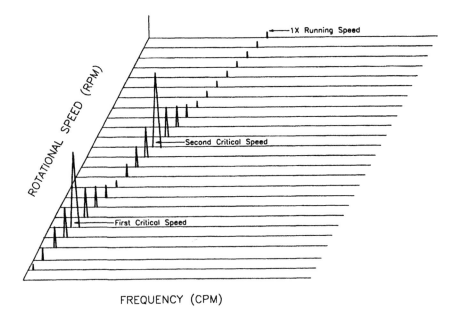

Figure 21.1 Campbell diagram.

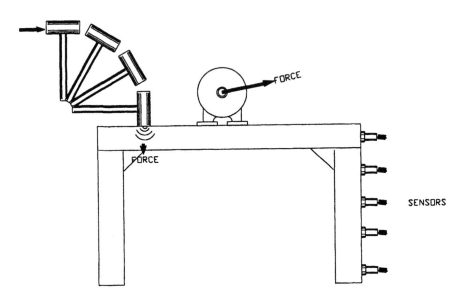

Figure 21.2 Simple machine support system.

result, it is time consuming to test each component. Unfortunately, this is the only positive means of isolating the offending component.

In a simple support system such as the one illustrated in Figure 21.2, the natural frequencies of the structure can be isolated by mounting multiple sensors on the structure and then hitting, or ringing, the structure with a zero-impact hammer. This approach is valid, but may not identify all natural frequencies of the structure. For example, the natural frequency of the right leg may not be the same as the left.

Therefore, the only accurate method to identify all natural frequencies is to isolate and ring or excite each structural member or static machine component. The methodology to be used depends on the type of machine component or structure to be tested and the source of excitation energy.

The excitation energy source can be difficult to determine, but in many cases it can be traced directly to one or more dynamic forces in proximity to the test structure. The possible sources of forcing functions, or excitation energies, include machine running speed(s), imbalance of a rotating or reciprocating element, misalignment, gear mesh, hydraulic/aerodynamic noise, and a variety of other abnormal dynamics that may be generated by machine-trains or process systems.

A number of energy sources can be used during testing to excite natural frequencies of stationary machine components or structural members. These sources include sinusoidal and nonsinusoidal vibration forces and ringing.

Sinusoidal Vibration Forces

Sinusoidal vibration forces can be used to excite the natural frequency of stationary components. However, these forces must be swept through the frequency range until they match the natural frequencies of the structure being tested.

Nonsinusoidal Vibration Forces

A nonsinusoidal force generates orders of the fundamental forcing frequency that, in turn, excites the structure's natural frequency. This phenomenon often occurs in normal machine operation.

Sometimes a shaker is used in conjunction with a power amplifier and a wave generator to excite the natural frequencies. The natural frequencies are determined by sweeping through the frequency range of interest. An example of a mechanical shaker is a variable-speed motor having a double-ended shaft with offset disks for mass unbalance. Other excitation sources, including random forces, can be added to the shaker to excite natural frequencies instantaneously.

Some frequencies generated by random forces coincide with, and thus excite, the machine-train's natural frequencies. This phenomenon can be seen in signatures taken from the operating machines where random energy in the system can excite the natural frequencies.

Ringing

The most popular method for exciting natural frequencies is to strike a machine or structure with a timber, or hammer. The range of frequencies excited depends on the duration of the impact.

Hard-faced steel hammers tend to bounce off a structure, thus providing a short-duration impact. As a result, this device only excites high frequencies. To excite natural frequencies of 10 Hz or less, a soft tip must be used on the hammer.

Isolating the Forcing Function

The preceding sections describe how to identify specific machine components or structures that are experiencing static resonance. In addition, they discuss how to isolate the unique natural frequencies. The next step is to find the sources of the forcing functions or energy sources.

The first step is to define clearly the specific frequency being excited. There must be an energy source with a frequency identical to the resonant frequency or some frequency within its generated broadband energy. The source of excitation energy can be determined either by calculation or by mapping.

Calculation

An analyst can easily calculate certain unique frequencies generated by a machine-train or process system. In many cases, the excitation energy source will be within the same machine-train or in proximity to the point of resonance. Therefore, the analyst should start with the machinery closest to the point of resonance. The following are examples of easily determined unique frequencies for a centrifugal pump rotating at 1800 rpm and having 10 vanes on the impeller:

- Fundamental frequency is equal to the rotating speed, or 1800 rpm.
- Vane-pass frequency (cycles per minute) is equal to the number of vanes on the impeller multiplied by the rotating speed (i.e., 10 vanes × 1800 rpm = 18,000 cpm).

While the calculation method does not confirm the location of an energy source, it provides a list of most likely sources of excitation. Direct measurement of these sources using a vibration analyzer can then be used to isolate the forcing function.

Mapping

Since the specific resonant frequency is known, the analyzer can be used to track the source of that unique frequency. If the excitation source is within the machine, the meter can be used to record the amplitude of the resonant frequency at various points around the machine. When a source of that specific frequency is found, the machine component adjacent to that measurement location is a probable source of excitation.

The same approach can be used to locate sources of excitation energy outside the resonant machine or structural member. Data can be acquired at regular intervals around the resonant member. If an energy source that coincides with the resonant frequency is observed, it can be tracked to its origination point using this method.

Natural frequencies are not always excited by an energy source that is a unique, concurrent frequency. Broadband energy sources also can be the source of excitation. For example, an unbalanced motor may not generate a measurable frequency that coincides with the observed resonance, but its broadband output may contain energy that coincides with, or is an integer multiple of, the natural frequency. Therefore, any high broadband energy is a potential source of excitation.

Testing Conditions

For resonance testing, the structure, piping, or machine should be as close as possible to its normal operating state. Parts of a machine cannot arbitrarily be removed and tested. For example, the natural frequencies of a gear that is not mounted on its shaft differ from those of a mounted gear. Similarly, the natural frequencies of a machine that is mounted for shop testing differ from those of a machine mounted on its normal foundation.

The level of sophistication and detail of resonance testing varies. A simple resonance test often provides the necessary structural natural frequency information. However, better information and greater detail can be obtained from more sophisticated tests and instrumentation.

Test Equipment

Tunable filter analyzers and storage oscilloscopes (analog or digital) can be used to observe the instantaneous vibration resulting from an impact or excitation of a structure.

A trigger level must be set on the oscilloscope in the single-sweep mode. The vibration signal from the impact will be held on the screen of the oscilloscope. The natural period is determined by counting the divisions in one period of vibration and multiplying that number by the time-per-division setting on the time base. The procedure using a digital oscilloscope is similar, except that the period is read directly with a cursor.

Rough estimates of the natural frequencies of a structure can be obtained with a tunable filter analyzer in the filter-out mode by observing the frequency meter after impact. The frequency meter will indicate the natural frequency as long as the structure is ringing.

The described methods evaluate the vibration response to a single impact. When such a test is conducted, the sweep filter is started and the structure is repeatedly bumped until the entire frequency range is scanned. Note, however, that the decay time for each bump is so rapid that they do not interact. The analyzer responds as the filter sweeps through the natural frequencies and the peaks of the envelope of the responses indicate the natural frequencies.

This bump test also can be performed using a FFT analyzer and a single impact rather than a series of impacts. In this case, the trigger on the analyzer is set to respond to impact from the hammer.

DYNAMIC RESONANCE

Rather than a Hanning window, the setting should be a uniform window. A variety of test methods can be used to identify dynamic resonance. Each has proven capability for its specific application, but generally cannot be used in other applications.

Constant-Speed Machines

This section reviews the best techniques for most common applications.

Constant-speed machines that are being operated at their design speed and load should not be affected by dynamic resonance. In most cases, dynamic resonance problems in this class of machine result from a radical change in the operating envelope (i.e., speed, load, etc.) or a modification of the machine-train.

Identifying dynamic resonance in a constant-speed machine-train is sometimes difficult. Routine monitoring such as that conducted as part of a predictive maintenance program detects the abnormal vibration levels that result from dynamic resonance, but do not clearly isolate resonance as the source of the problem.

In the previous fan example, the fundamental running speed of the fan shaft is the forcing function and the first critical speed is the resonant frequency. Both appear at the running-speed frequency. Because imbalance and most other failure modes also result in an increase in the fundamental (1x) frequency component, the question is how to separate these failure modes from resonance.

The major difference between resonance and other failure modes is the amplitude of the frequency component. Normally, common failure modes such as imbalance and misalignment increase the amplitude of the fundamental (1x) frequency, but the increase is small to moderate when compared to prior readings or to other frequency components.

Dynamic resonance will dramatically increase the amplitude of the natural frequency component. Typically, the relationship between the energy levels at the resonant or natural frequency is an order of magnitude or more higher than any of the normal rotational frequencies associated with the machine-train. In addition, the resonance frequency tends to have a broader base than normal rotational frequencies with no modulations.

The only positive means of isolating resonance in a constant-speed machine is to perform a coast-down or run-up transient analysis. This type of analysis can be conducted with the vibration analyzer in the transient-capture mode. Mounting transducers and a tachometer on the suspect machine, then recording the amplitude and phase of the suspected resonant frequency performs the test.

In a coast-down test, the machine-train's driver component is turned off and the suspect frequency is recorded as the speed decreases. If the problem is critical speed, the amplitude decreases rapidly as the speed decreases. In addition, there is a 180-degree phase shift as the machine's speed passes through the resonant zone.

A run-up test uses the same setup, but records the amplitude and phase as the machine accelerates from dead-stop to full speed. The amplitude of the natural frequency dramatically increases as the speed coincides with the rotor's natural frequency. In addition, the natural frequency's phase shifts 90 degrees as the machine's speed enters the resonant zone. If the speed continues to increase and leaves the upper limit of the resonant zone, the phase will shift an additional 90 degrees.

Variable-Speed Machines

Isolation of resonance in variable-speed machines (e.g., cold reduction mills) can be accomplished by acquiring a waterfall plot of signatures during either run up or coast down of the machine (see Figure 21.3). The waterfall plot consists of a series of frequency-domain vibration signatures that are captured periodically throughout these two transients. The frequency of data capture can be controlled by either a tachometer input or time interval, but should be high enough to ensure complete coverage of the acceleration and deceleration of the machine-train.

Typically, signatures should be captured at least once per second or every five revolutions of the primary drive-shaft. This frequency is, in most cases, sufficient to capture the transition of the running speed through any natural frequencies or resonant zones.

The waterfall plot clearly displays resonant zones that are crossed by any of the rotational frequencies within the machine-train. If the resonance is excited by the major components (e.g., running speed, gear mesh), the waterfall plot will, in most cases, clearly display the resonance as a major increase in amplitude of the resonant frequency.

For example, the gear-mesh energy from a gearbox or a pinion stand in a cold-reduction mill typically passes through one or more resonant zones as the mill accelerates or decelerates. The transients are displayed as momentary increases in the amplitude

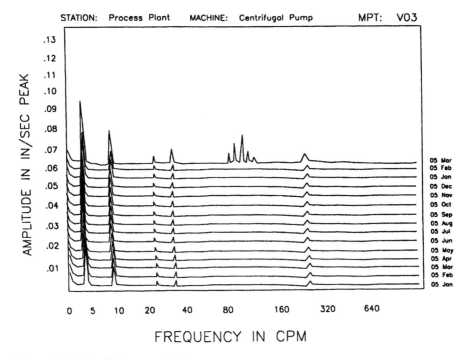

STATION: Process Plant   MACHINE: Centrifugal Pump   MPT: VO3

Figure 21.3 Waterfall or cascade plot.

of the gear-mesh frequency as it passes through the resonant zone. In the same fashion, the running speed of the mill also passes through resonant zones during acceleration and deceleration.

Most variable-speed machines exhibit this same type of momentary resonance. Normally, it is not harmful and does not require corrective action. As long as the transition through the resonant zone is made quickly and smoothly, there is no adverse effect on the machine-train. However, if the excitation force (i.e., running speed or gear mesh) is maintained for some time within a resonant zone, serious damage may occur. Severe chatter or even catastrophic failure of the mill or one of its components, such as a roll, could occur.

Dynamic resonance is not always driven by an obvious excitation source. Because extremely low levels of vibration energy can excite a natural frequency, other machine components, such as bearings, can be the source of extremely destructive resonance. Using the cold mill as an example, one of the most common sources of resonance is the outer-race passing frequency of a work-roll bearing. Because of roll-bending practices, the load zone in the work-roll bearing is shifted to the outer race. The low-level energy created by the balls or rollers passing the outer race is sufficient to excite the natural frequency of the work roll. When this occurs, the resonance of the roll results in roll bounce or eccentric rotation. In turn, this bounce causes gauge deviation in the strip, or chatter.

The most important point to remember when searching for the excitation source of resonance is that the amplitude or strength of the source does not have to be high. Amplitudes of less than 0.03 IPS-PK can be enough to excite the rolls or mill stands and result in a measurable gauge deviation in the strip.

Chapter 22

MODE SHAPE

Machinery rotor assemblies, such as shafts, are designed to rotate on their true center-line and in a state of equilibrium. In practical applications, however, few machines achieve optimum or design conditions. As a result, most machines operate with a slight amount of imbalance where the shaft rotates slightly off of its true centerline. Such rotor deflection is referred to as mode shape.

Even in a perfectly installed system, imbalance often results from the difference between the lift that was designed into the rotor assembly and the force of gravity that pulls the rotor downward. Because the effects of gravity vary with altitude, and the lift varies with temperature, humidity, and barometric pressure, there is always some dif-ferential. As a result, all machine trains have some imbalance and rotate offset from their true centerline. The slightly eccentric or off-center rotation of the shaft generates a low-level frequency component that coincides with the rotating speed of the shaft. This type of rotor dynamics is called the first mode.

In addition, all rotating shafts are flexible and change shape throughout their normal operating range. The shape or shapes a shaft takes in actual operation are a function of shaft stiffness, bearing span, rotor weight, and the action of outside forces such as aerodynamic or hydrodynamic forces generated by the system. In ideal systems, the shaft retains a relatively straight shape, but does not rotate on its true centerline. This offset or eccentric rotation is one form of the first mode of the rotor assembly. If the rotor shaft deforms and bends, but maintains its original node points (refer back to Figure 19.3), it will continue to generate a vibration frequency at the fundamental running speed of the shaft (1×). The amplitude of this fundamental frequency is directly proportional to the amount of bend or deflection between the shaft bearings.

If the shaft flexes into a double bend that crosses its true centerline, it creates a node point at the point at which it crosses the neutral or centerline. As the shaft rotates, the double-bend shape creates two high spots as it passes the vibration transducer. These high spots are interpreted as the fundamental (1×) and second-harmonic (2×) frequencies of running speed. This profile describes the second mode of the rotor assembly.

In some applications, the shaft can flex or deform into mode shapes that generate third (3×) and fourth (4×) running-speed harmonics. Cantilevered rotating machine elements, such as on overhung fans, are typical examples of machinery subject to shaft deflections that force operation in the third or fourth mode.

Part IV

REAL-TIME ANALYSIS

Chapter 23

OVERVIEW

Real-time analysis (RTA) is an advanced diagnostic technique. It is especially useful with complex machinery, for evaluating transient events, such as rapid speed changes, or where the steady-state vibration data gathered by conventional vibration analyzers are not sufficient to isolate the root cause of a problem.

HARDWARE REQUIREMENTS

Real-time vibration analyzers provide many benefits that are essential for a comprehensive vibration-monitoring and analysis program. They provide the ability to analyze a dynamic signal and they permit the analyst to view the actual dynamics of a machine or process system in real time. As a result, the analyst can evaluate the actual cause-and-effect relationship of variables, such as load, speed, and product parameters, as changes take place. This unique ability permits the analyst to isolate specific changes in the machine's or system's operating envelope that directly affect the operating condition, reliability, and useful life of the machine or system.

The diagnostic logic used for RTA is based on the same concepts as traditional time- and frequency-domain vibration analysis. However, it provides the ability to view the interactions of the machine's vibration components dynamically rather than statically. This greatly improves analysis accuracy and the ability to isolate the true root cause of problems. As illustrated in Figure 23.1, real-time analyzers process and display multiple data formats in real time.

A real-time vibration analyzer is essentially a conventional microprocessor-based vibration analyzer with expanded capability. These analyzers acquire a time-domain data signal and convert it to frequency-domain by performing a fast Fourier transform (FFT) on the data. Both conventional and real-time instruments have filters and sig-

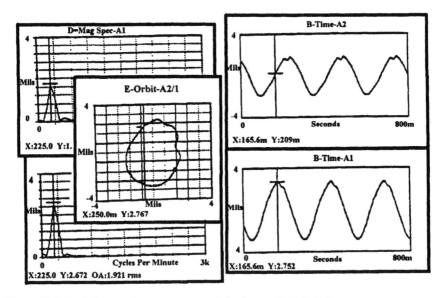

Figure 23.1 Real-time analyzers process and display multiple data formats.

nal-conditioning logic that convert the mechanical forces within the machine-train into vibration data that are displayed in units, such as inches per second peak (i.e., velocity data), mils peak-to-peak (i.e., displacement data), or *g*'s (i.e., acceleration data). Most of these instruments provide easy data acquisition and display of frequency-domain vibration signatures.

The major difference between real-time analyzers and routine vibration-monitoring instruments is processing speed. A real-time analyzer can evaluate complex vibrations and noise hundreds of times faster than conventional analyzers. These instruments process the acquired raw signal in a fraction of the time that it would take a general-purpose instrument.

Additional data ports allow real-time analyzers to acquire and process simultaneous channels of data, which is a major advantage over conventional analyzers. Unlike conventional vibration-monitoring instruments, real-time analyzers have multiple, independent circuits that can process incoming data in parallel. They include synchronizing logic to ensure that all channels of recorded data are real time and absolutely parallel. The ability to evaluate multiple channels of data that are absolutely synchronized in real time is a major advantage of this diagnostic technology. This capability permits the analyst to evaluate the interactions of the entire machine-train, or the machine-train and its installed system. Because the operating dynamics of a machine are dependent on these interactions, this type of analysis (i.e., multichannel analysis) is an invaluable diagnostic tool.

LIMITATIONS OF MICROPROCESSOR-BASED DATA COLLECTORS

Although real-time microprocessor-based, vibration-monitoring instruments are less limited in their functional capabilities than conventional instruments, they share some limitations, but are designed to overcome others. Conventional microprocessor-based, frequency-domain data collectors are designed to acquire only a single channel of vibration data while real-time analyzers acquire multichannel data. As a result, conventional units cannot provide all of the capabilities necessary for all diagnostic applications. This limitation is discussed later as are others, including signal-conditioning methods (i.e., bandwidth limitations, spurious signal rejection logic, and antialiasing logic), display types, data-processing speed, and "near" real-time data.

Single-Channel Data Limitations

For more complex machine-trains or in applications where speed, load, or other process variables change frequently, the use of single-channel data collection has limited value. For example, a variable-speed machine may alter speed during the data-acquisition process. If this occurs, the single-channel data collector will either smear the fundamental frequency or delete the blocks of data that contain the speed change. Either of these actions will distort the vibration profile of the machine-train and severely restrict the analyst's ability to evaluate accurately the machine's operating condition.

Although there are many applications where single-channel, frequency-domain data acquisition is insufficient, it is acceptable for relatively steady-state operating conditions, such as those found with constant-speed motors, pumps, and other simple machine-trains. In these cases, even though the machine-train data are acquired in series (i.e., one point at a time with some time lag between readings), the total data set provides a reasonably accurate profile of the machine-train's operating condition.

Signal-Conditioning Methods

All microprocessor-based data collectors contain signal-conditioning logic designed to ensure accuracy of acquired data. While these filters and the logic used to condition the input signal are valid for relatively steady-state machine-trains, they can distort data from variable-speed/variable-load machines. In particular, three signal-conditioning logic factors can cause problems: bandwidth limitations, spurious signal rejection, and antialiasing logic.

Bandwidth Limitations

The signal logic used to condition the incoming signal from a transducer limits the bandwidths that can be used for data collection. While microprocessor-based instruments permit a select range of maximum frequencies, F_{MAX}, the minimum frequency limit, F_{MIN}, for signal conditioning is always zero. This limitation is not obvious on some systems because, although the software permits the user to select a F_{MIN} other than zero, the input value is not used during the data-acquisition sequence.

The disadvantage of a fixed F_{MIN} of zero is that it prevents acquisition of a high-resolution vibration profile centered around a unique frequency band. This ability, called zoom, is available in most real-time analyzers and is a necessary diagnostic tool for complex machinery.

Spurious Signal Rejection Logic

Signal-conditioning logic eliminates periodic vibration frequencies that do not repeat within each block of data acquired by the microprocessor. This logic is used to eliminate spurious frequencies, such as impacts, transients, or electronic noise, that might distort the vibration signature. The disadvantage of such logic is that failure modes due to such events cannot be detected when using this data-acquisition technique.

The microprocessor's signal-rejection logic automatically evaluates each block of data as it is acquired by directly comparing it to preceding blocks. If the rejection logic detects frequencies in the new block of data that were not present in the preceding block, it automatically rejects the new block. Because the microprocessor automatically acquires an additional block of data until each of the acquired blocks contains nearly identical frequencies, the technician may be unaware that this is occurring. In extreme cases, the instrument will abort the data-acquisition process due to radical, repeated fluctuations in speed, load, or other variables that change either the vibration frequencies or amplitude.

Antialiasing Logic

Most single-channel vibration instruments include an antialiasing filter in their signal-conditioning logic. These filters are designed to prevent vibration signature distortion caused by fold-over of both low- and high-frequency components. Fold-over is where high frequencies invert or "fold" and appear as higher frequencies. While prevention of distortion is important, the use of preset filters can limit the diagnostic capability of these instruments.

Display Type

Conventional microprocessor-based systems used for vibration monitoring in traditional predictive maintenance programs acquire, condition, and store an averaged snapshot of the machine's vibration profile. This snapshot is later downloaded to a desktop computer for analysis. This snapshot view severely limits the analyst's ability to understand the dynamics of the machine-train because the vibrations generated by a machine are not static, but are dynamic.

Both the overall vibration energy and that of each frequency component are constantly changing. The phenomenon, referred to as beats or beating, is indicative of the machine's operating condition and can provide a powerful tool for analysis. However, data-processing speed, display logic, and diagnostic logic used in conventional vibration-monitoring instruments prevent them from being used to monitor this phenomenon.

Real-time analyzers are required to take advantage of the data provided by the phenomenon. These analyzers permit direct or indirect data acquisition and provide a dynamic display of the time-domain waveform, frequency-domain signature, or both. Dynamic data allow the analyst to better view and interpret the condition of the machine-train. Dynamic displays show the interaction of the various frequencies that make up a machine's signature. This is especially important when two or more frequencies couple, which can significantly influence the overall signature.

Data-Processing Speed

Most conventional data-acquisition instruments that are used for routine vibration monitoring and analysis do not have the data-processing speed needed to perform RTA.

Chapter 24

APPLICATIONS

There are many potential applications for real-time analysis, ranging from advanced vibration analysis to structural and process analyses. However, the major applications for this diagnostic technique are transient analysis, complex-machine analysis using synchronous time averaging and narrowband zoom, multichannel analysis, and torsional analysis.

TRANSIENT ANALYSIS

Many of the machines and process systems found in manufacturing and process plants are periodically subjected to events that directly affect their operating dynamics, reliability, and useful life. These events, called transients, may be part of the normal operating mode or they may be an external incident or variable not uniquely associated with the machine or system. Specific applications that require transient analysis include variable-speed machines, process or product variations, and random or periodic impacts.

A transient is a short-duration event that periodically occurs in a machine or process system. As an example, when a variable-speed machine's running speed coincides with its rotor's critical speed, that event is considered to be a transient. In normal practice, the duration of this type of event is—or should be—relatively short. Because they generally occur so quickly, conventional vibration-monitoring instruments do not have the processing speed required to capture the impact of the event and, therefore, a real-time analyzer is required to perform the proper diagnostics.

Figure 24.1(a) illustrates a transient captured in a time waveform and Figure 24.1(b) illustrates the same transient in a waterfall or cascade frequency-domain plot. These capture-and-display capabilities account for much of the diagnostic power of RTA.

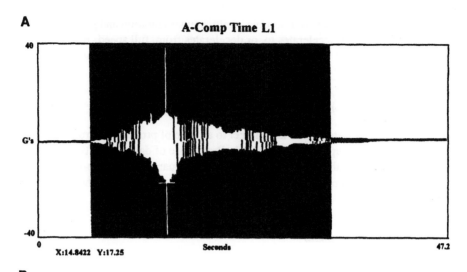

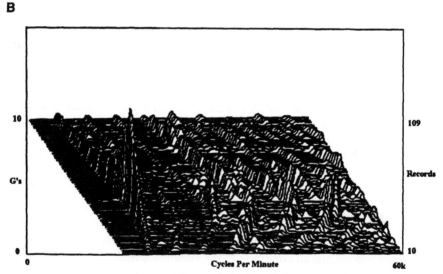

Figure 24.1 (a) Time waveform of transient event. (b) Same transient in waterfall format.

Variable-Speed Machines

Paper machines and steam turbines are examples of variable-speed machines that operate over a wide range of speeds and are excellent applications for RTA. Because the dynamics of the machine and its components change during the transitional period associated with any substantial speed change, the ability to measure and evaluate these changes is critical to any reliability program.

RTA permits accurate capture of real-time, multichannel vibration and process parameter data as the machine accelerates to, or decelerates from, full speed. Because these variations are rapid, the increased data-processing speed of a real-time analyzer is essential.

Process or Product Variations

Machines or process systems that are subject to a range of product- or process-related variables are also ideal applications for RTA. Proper use of data-acquisition and analysis techniques permits the analyst to fully define and quantify the cause-and-effect of the variables on the operating dynamics, reliability, and useful life of the machine or system component.

Random or Periodic Impacts

Many of the machine-trains and process systems that make up a plant are subjected to apparently random or periodic impacts, or transients, that may affect their reliability or useful life. Conventional vibration monitoring will not detect these events except by accident.

The combination of digital tape recorders and RTA can be used to document and evaluate these occurrences. Using a digital tape recorder to acquire vibration and process data over a long period of time is a cost-effective way to capture those random or long-interval periodic events that may affect a machine's reliability. When combined with the tremendous data-processing speed of a real-time analyzer, this provides the analyst with the means to evaluate these events quickly to isolate their root cause.

COMPLEX-MACHINE ANALYSIS

Not all machines or process systems within a plant are simple. Many are comprised of complex components that all contribute to the vibration waveform or signature. As a result, analysis of the vibration profile is often very difficult. For example, a triple-reduction gear box, coupled to a variable-speed steam turbine and variable-pitch fan, generates a very complex vibration profile. Each of the components in the driver, gearbox, and fan contributes to the time waveform or frequency-domain signature. Conventional vibration meters cannot separate the individual contributions of these components from the composite profile.

Real-time analyzers can be used to break up the individual frequencies generated by various machine or system components because of its data-processing speed and signal-conditioning logic. Two RTA techniques are especially useful for complex machines: synchronous time averaging and narrowband zoom.

Synchronous Time Averaging

When synchronous time averaging is coupled with the real-time analyzer's ability to provide parallel, multiple data channels and real-time processing, RTA can be used to evaluate extremely complex vibration profiles. This technique removes unwanted frequencies that may be generated by other components either within the same machine, from other machines, or from unknown outside sources. Simply stated, this technique nullifies or removes any vibration frequencies that are not absolutely synchronous to the shaft or turning speed being investigated.

Narrowband Zoom

One of the limitations of conventional, general-purpose vibration instruments is their inability to provide high-resolution signatures of specific frequency components (e.g., gear-mesh, vane-pass, or bearing modulation). Because these systems have a fixed lower data acquisition frequency limit (i.e., zero), they cannot provide resolution on any set of frequencies that has a lower limit greater than zero. As a result, the analyst may not be able to visually fully evaluate the profile generated by a specific machine component.

Most real-time analyzers can provide a true high-resolution signature for any set of frequency components within the machine vibration profile. This ability, called narrowband zoom, is a powerful diagnostic tool. It is especially useful for complex machines where multiple components may require close evaluation. As an example, narrowband zoom can be used to radically increase the resolution of the gear-mesh frequency of a gear set. The analyst can establish a narrowband-zoom window by selecting the gear-mesh as the center frequency and defining a bandwidth that includes the gear's modulation frequencies. The resultant display provides a high-resolution signature that can be used to fully analyze the gear's condition.

MULTICHANNEL ANALYSIS

Multichannel analysis is not unique to RTA. It can be performed with conventional, general-purpose vibration instruments using a digital tape recorder and good discipline. However, this approach is more difficult and does not provide the diagnostic power of RTA.

Conventional vibration-monitoring systems are limited by their processing speed and functional capabilities. Therefore, they cannot provide all of the tools needed for full multichannel analysis. In addition, using them requires the raw data to be acquired using multichannel tape recorders and then played back through the general-purpose meter. As a result, playback timing becomes a real problem. There is no guarantee that all data are time synchronized or that the displayed data are meaningful.

Use of a real-time analyzer, with or without a digital tape recorder, eliminates these problems. The multichannel, parallel-processing capabilities of the analyzer provide a quick, positive means of retrieving and displaying data that are absolutely time synchronized.

TORSIONAL ANALYSIS

Torsional vibration of rotating elements is the rapid fluctuation of angular shaft velocity, and its basic units are either radians or degrees. A machine will often increase or decrease speed over some period of weeks, days, or seconds. As a machine changes speed, torque is applied to the shaft in one direction or the other.

Torsional vibration is not a simple parameter to analyze. Transducer requirements are stringent and shaft access may be limited. Above all, however, there is a peculiar mystique engulfing torsional vibration. Therefore, this module attempts to dispel its mystique by providing a basic understanding of torsional motion, what it means, and how it can be interpreted.

Chapter 25

DATA ACQUISITION

This section provides the basic information needed to acquire accurate real-time data. It assumes that the analyst or technician is familiar with microprocessor-based real-time spectrum analyzers, digital tape recorders, and other appropriate instrumentation. The users' manuals for the actual instruments to be used should be consulted in conjunction with this training module.

Regardless of whether direct or taped data acquisition is selected, the approach used to gather real-time data is the same as for single-channel (i.e., route) acquisition. The same rules are used for measurement point location and orientation, analysis parameter set selection, measurement point definition, etc. The only exception is that all data are broadband, rather than both broadband and narrowband.

Before using a real-time analyzer as part of the periodic monitoring program, the technician or analyst should review the instructions provided for data acquisition. All of the rules and methods used in the routine monitoring program apply to real-time data acquisition. In addition, he should thoroughly review the users' manuals for all other instruments to be used for data acquisition and analysis. However, recently purchased real-time spectrum analyzers use a Microsoft Windows-based operating system, which greatly simplifies their use. Like a personal computer, all functions of the analyzer can be accessed from the main menu using standard Windows protocol.

Input for all data fields on the acquisition setup must be included for all active channels before attempting data acquisition. Care must be taken to ensure that all data are consistent. Unlike a single-channel system, a real-time analyzer provides exactly what is requested. If errors or inconsistencies are made in the acquisition setup, it will perform the preprogrammed statistical or mathematical functions. It will not question errors or inconsistent formats between the various data fields.

For example, if the user selects acceleration data (e.g., 100 mV/g) as the calibration factor and velocity units of "inches per second" as the engineering units name, the real-time analyzer will acquire and display the vibration data as velocity readings even though it has not integrated the acceleration data into velocity. As a result, the displayed data will have no value as a diagnostic tool.

METHODS

There are two ways to acquire the data needed to perform a RTA: direct acquisition and tape recording.

Direct Acquisition

In direct-acquisition mode, the real-time analyzer can be used to acquire a variety of vibration and nonvibration, process system data, which are stored directly in its on-board memory. Therefore, the primary limitations of direct acquisition are the on-board memory capacity and the inflexibility of the stored data.

The advantage of direct acquisition is that monitoring of machine-train or process system operating conditions can occur as the data are acquired. This allows the analyst to adjust the data-acquisition parameters as needed to ensure accuracy. In applications where a quick diagnosis is needed, this approach provides a means of isolating and solving simple problems.

The disadvantage is that it extends the time and manpower required. Unlike the pre-programmed, microprocessor-based analyzers used for routine vibration monitoring, real-time analyzers must be manually configured for the specific type of data before each acquisition. For example, it can either be configured to acquire time waveforms, frequency-domain signatures, high-resolution narrowbands, or a variety of others. The analyzer acquires, conditions, and displays a continuous profile in the user-selected format. If the analyst wants to look at a different data format, she must abort the data acquisition and reset the instrument for the new data format. In addition, the acquired time- and frequency-domain data are not time synchronized, but are taken in series. Data taken in series eliminate the ability to compare the time trace with the frequency-domain signature.

However, the biggest disadvantage of the real-time analyzer is that data, once captured, cannot be converted to a different format. For example, time traces cannot be converted to frequency-domain data. This limitation can greatly restrict the diagnostic capability of the analyzer or increase the analysis time where multiple data formats are required for proper analysis. Because the user cannot view real-time data in more than one format, he must reacquire it each time a different format is required. The problem with this technique is that each data set is new. As a result, subsequent data-acquisition runs may not duplicate transients or the operating condition of the machine-train found in a previous run.

Tape-Recorded Data

With this approach, the analyst uses a tape recorder to acquire the data, which is captured and stored on tape. This approach permits quicker acquisition of data that can be analyzed quickly in parallel, or an in-depth analysis of the machine-train or process condition can be performed at a later date.

Types of Recorders

Two major types of tape recorders are used to acquire vibration and process parameter data: analog and digital. Each type has advantages and disadvantages that should be understood before using them for RTA. The major difference between the two types of recorders is that, while they both take an analog signal as input, the digital recorder incorporates an analog-to-digital converter. This is a device that translates continuous analog signals into proportional discrete digital signals.

Analog Recorder

Analog signals are nominally continuous electrical signals that vary in amplitude or frequency in response to changes in sound, light, heat, position, or pressure. Analog recording is any method in which some characteristic of the recording signal, such as amplitude or frequency, is continuously varied in a manner analogous to the time variations of the input signal. The two major types of analog tape recorders used to acquire vibration and process parameter data are direct-record and frequency-modulated units. The major difference between these devices is in their ability to record low-frequency signals.

Direct-Record Tape Recorder

With direct-record analog units, the signal amplitude is captured directly by the tape's magnetic field. Therefore, variations in tape quality and ambient conditions (i.e., heat, light, and stray magnetic fields) directly affect the data obtained with this type of recorder. This type of device cannot record frequencies below 25 Hz, or 1500 rpm. This is because playback is based on the rate of change of tape magnetization.

Frequency-Modulated Tape Recorder

With frequency-modulated analog units, the signal amplitude is recorded as the difference between a base or carrier frequency and the frequency recorded. As a result, the frequency-modulated recorder is much less sensitive to variations in ambient conditions and the magnetic properties of the tape used for data acquisition. Frequency-modulated recording can be used with low frequencies down to the physical limits of the transducer, signal conditioning, and cable that are used.

Digital Recorder

Digital recorders have the ability to condition and filter the raw input signal in much the same way as single-channel vibration analyzers and multichannel real-time analyzers. In this type of tape recorder, the incoming signal is passed through an analog-to-digital converter and stored in a digital medium as a series of digital values. Most of these instruments can filter the analog data to prevent aliasing and to condition the output to user-selected values.

Recording the Data

Because it is difficult to anticipate the exact formats and data that will be required to resolve a machine-train or process problem, full-range tape recording of data is the recommended practice. Storing the data on tape ensures that the raw data will be available for complete, comprehensive analysis.

Data-Acquisition Practices

Unlike vibration data that are collected with traditional microprocessor-based predictive maintenance programs, real-time data collection does not use preprogrammed acquisition routes. Therefore, the acquisition route for obtaining each data set must be set up and performed manually. As a result, acquiring this type of data requires more time, discipline, and expertise than for routine vibration monitoring.

The following sections discuss the practices that should be followed to ensure that accurate, meaningful data are obtained. In particular, the following topics are discussed: hardware setup for transducers, cables, and power supplies; channel integrity; test plans; and field notes for channel data, transducer data, gain, and sequence of events.

Hardware Setup
RTA is generally used in conjunction with multichannel data acquisition, which complicates the hardware setup requirements. Therefore, the required hardware setup is quite different than that used for routine vibration monitoring. This section discusses the setup requirements for the transducers, cables, and power supplies that are needed.

Transducers
Transducers, which are used to obtain vibration or process data, must be selected with care. In particular, they should be compatible with the specific measurement parameters of an analysis. Generally, accelerometers should be used to acquire the vibration data for a RTA. This type of transducer is better suited for most applications because it is less sensitive to mechanical damage and temperature.

The accelerometers should be of the low-mass variety and have a positive means of mounting to the machine-train (e.g., stud, epoxy, or magnet). In addition, they must have the linear-response characteristics needed for the specific application. Each accelerometer should have a certified specification sheet that defines its operating range and response characteristics. It also should have a current calibration test.

Cables
Unlike general-purpose vibration monitoring, RTA typically requires massive cable runs to connect the multiple channels to a digital or analog tape recorder, or directly to the real-time analyzer. Both the number of cables and the average run length create

unique problems with this type of analysis. Generally, two types of cable are used for a RTA: microdot and coaxial.

Microdot cables are normally required to make the initial connection between a low-mass accelerometer, power supply, and tape recorder or analyzer. The cable is a small-diameter (i.e., about 1/16 in.) assembly that includes threaded connections. Because of its size, microdot cable is extremely sensitive to misuse or physical damage. Therefore, care must be taken to ensure that it is protected throughout the data-acquisition sequence.

The use of microdot cable assemblies should be minimized as much as possible. In addition to their sensitivity to damage, the resistance within the cable may distort the electrical signal. Wherever possible, total microdot runs should be less than 5 ft. Longer runs may cause attenuation or distortion of the signal.

Coaxial cables are used for the long runs that connect the transducer to either a tape recorder or real-time analyzer. These cables have a larger diameter than microdot cable and are almost immune to damage. They are similar to those used for cable television connections and provide a reasonably reliable way to make critical connections.

Total runs between the transducer and recorder should not exceed 70 ft. Signal attenuation beyond this distance has a severe effect on data quality. If longer runs are required, a signal amplifier can be added to each cable to boost the signal strength and permit the longer run.

Power Supplies

All transducers require a power source to operate properly. In general-purpose vibration monitoring, the power source is usually part of the analyzer. In many real-time applications, however, an external power supply must be provided for each accelerometer or transducer.

The external power supply must be matched to the transducer. For example, most accelerometers require a 4-mV power supply to function properly. In addition to their compatible rating, power supplies must provide constant, reliable power throughout the data-acquisition sequence. Because many of the power supplies that are normally used in this type of application are battery powered, care must be taken to ensure that fresh batteries are installed at the beginning of each data-acquisition sequence.

Many power supplies include an amplifier, or gain, that can be used to increase the raw signal strength of the transducer. While this ability is helpful with weak signals, it can lead to serious diagnostic errors. Typically, the gains provided by power supplies are in steps of 10, ranging from 0 to 100. For example, if the user selects an amplification factor of 100×, the signal strength recorded by the analyzer will be 100 times higher than the actual vibration energy.

Channel Integrity

In all RTA applications, extreme care must be taken to ensure data accuracy. This is especially true when the analysis is combined with multichannel data-collection techniques. It is imperative for the analyst to be able to identify absolutely each of the channels as data are acquired.

Permanently numbering components used for each data-acquisition channel is the best assurance of this ability. Everything from the accelerometer to the final connection on the coaxial cable should be numbered. Permanently affixed cable tags should be on both ends of all cable assemblies, as well as other channel components.

The entire cable run for each channel should be inspected and verified prior to a data-acquisition sequence. In addition, a continuity test should be conducted on each channel to ensure a distortion-free channel.

Test Plans

Applications that require RTA techniques are generally more complex than those that are appropriate for traditional vibration monitoring and analysis. Typically, RTA is used for complex applications, such as torsional problems, and a series of well-planned data acquisitions and analyses are required. Therefore, a detailed test plan is essential.

The test plan should concisely define the specific tests that will be performed. For each of these tests, the plan should include the setup data that will be needed to install and connect the transducers, power supplies, cables, and other instruments.

Field Notes

The analyst must document the exact events, timing, and data-acquisition methods used to record the information for each data channel. Because analysis may take place at some time after data acquisition, the analyst must have sufficient documentation to fully understand exactly when, where, and how the data were recorded.

Many of the digital tape recorders provide a voice-over channel that permits direct verbal commentary that can be played back during analysis. However, detailed, written notes also are essential. Documentation should include the following: channel data, transducer data, gain, and sequence of events.

Channel Data

The test log for each data set should clearly identify the location and orientation of each transducer. This information should be verified during the data-acquisition sequence to make sure that it is accurately recorded.

Transducer Data

The test log should include the specific type and setup of each transducer used for data acquisition. As a minimum, the log should include model number, serial number, and engineering conversion unit (i.e., 500 mV/g, 1000 mV/g, etc.).

Gain

In most cases, an external power supply or signal-conditioning instrument is used in conjunction with the transducers. Both the power supply and signal-conditioning units have the capability, called gain, to increase the strength of the raw signal. For example, a typical gain from a power supply is 10x. When this setting is selected, the raw signal strength is increased by a factor of 10.

The gain that is used must be recorded so the analyst can accurately evaluate signal strength. If the analyst is unaware of the actual gain, she will believe that the signal strength is 10 times higher than the actual value.

Sequence of Events

Included in the documentation needed to define the data set should be a concise description of the test, channels recorded, and the start-to-end timing of the data-acquisition process. The information should include all known variables and any assumptions that may have affected the data.

PARAMETERS

Most analyzers have up to eight channels that can be used for data acquisition. Each of the active channels to be used for data input, processing, and display must be set up manually at the beginning of each data-set analysis. Therefore, extreme care must be taken to ensure that all active channels are properly set up and that both the data-acquisition and data-analysis parameters are consistent. The parameters required for proper data acquisition include channel coupling, full-scale voltage, calibration factor, engineering units name, and trigger group.

Channel Coupling

Coupling is selected on a channel-by-channel basis and defines how the input signal is conditioned during the data-acquisition sequence. There are three choices for channel coupling: alternating current (ac), direct current (dc), and internal power supply.

Alternating Current

When the signal source is ac, the dc component is rejected and only the ac component is acquired by the analyzer. When real-time vibration data are to be acquired, this is the normal mode of signal conditioning.

This is not the case when the analyzer is used for direct acquisition of data. Selection of the ac-coupling mode will not provide power to the accelerometers or other transducers used as part of the direct-acquisition mode of operation. Therefore, the ac-coupling option should not be used for direct data acquisition unless external power sources are used to drive the transducers.

Direct Current

When the dc-coupling mode is selected, both the ac and dc components of the machine's vibration profile are acquired by the analyzer. In most cases, the dc component is comprised of electronic noise that distorts the vibration profile acquired from the machine-train. When a real-time analyzer is being used purely as a vibration analyzer, this option should not be selected.

Internal Power Supply

Many real-time analyzers have an internal power supply. Unless an external power source is used, this option should be selected for all direct data-acquisition applications. It provides a 4-mA/4-V dc power source that can be used to power a compatible accelerometer or other transducer.

This option should not be used when tape-recorded data are transferred into the analyzer. Transferring taped data to an analyzer requires an ac coupling.

Full-Scale Voltage

Unlike single-channel, microprocessor-based vibration analyzers, real-time analyzers do not automatically autoscale the input vibration signal to establish the maximum signal amplitude. Therefore, the user must select a maximum input voltage before acquiring data. The full-scale (FS) voltage option presets the maximum vibration level to be recorded by the analyzer.

The full-scale value, which is usually expressed as root mean square (RMS) must be selected on a channel-by-channel basis. Care must be exercised to ensure that selection for all channels is completed before acquiring data.

Most analyzers permit selection of an amplitude scale between 1 mV and 20 V set in increments of 1, 2, 5, or 10 mV. This range is more than adequate for most applications, but care must be taken to ensure that the input signal is not amplified above the FS voltage.

Care must be taken when selecting the FS RMS. Too low a value will "clip" the frequency components and not provide a true indication of the total amplitude of individual components or the overall, or broadband, energy represented by the data point. Loss of the actual amplitudes prevents proper analysis of the data and, hence, the machine-train's condition.

Most analyzers' autoscale function will not override the FS RMS scale selection in either the data-acquisition or analysis mode. When data are clipped by a low FS RMS selection, it cannot be recovered.

If the FS RMS scale is too high, it may exceed the analyzer's dynamic range. In this instance, the amplitude of the major frequency components is displayed, but the lower level frequency components may be lost in the noise floor. While most analyzers have a good dynamic range, the potential for masking important frequency components is high when the maximum FS RMS (20 volts) is selected.

Calibration Factor

The calibration factor is used by the real-time analyzer to convert channel voltage to the more convenient engineering units (EU). This option is used to convert the raw voltage reading (in millivolts) into more usable units of measurement, such as those for velocity, acceleration, or displacement.

The user must enter the appropriate calibration factor for the accelerometer, velocity transducer, or displacement transducer used to collect data. This conversion factor must be entered for both direct or tape-recorded data. In most cases, the conversion factor will be 100 mV/g for a general-purpose accelerometer, or 500 mV/g for a low-frequency accelerometer. However, the user must define the actual response characteristics of the transducer used in each application.

Vendors generally include certification curves and specification sheets for transducers, including accelerometers. This documentation, which should have been retained upon purchase, provides both the conversion factor and the response characteristics of the transducer. This information is required to perform a RTA.

Engineering Units Name

In routine vibration-monitoring equipment, the preprogrammed measurement routes include a conversion factor from raw input voltage (in millivolts) to a user-selected value, such as velocity, peak, or mils peak-to-peak, and do not require this parameter to be input.

Most real-time analyzers, however, do not offer this automatic conversion. The engineering unit (EU) name setup parameter identifies by name the type of unit (i.e., psi, mils, speed, etc.) that is needed for each data channel.

The EU name can be set using the standard keyboard in the same manner as the calibration factor. The analyzer will accept any string from one to six characters in length, but the units should be the same as for the calibration factor. For example, an accelerometer with 100 mV/g response should have an EU name of "g's" or "accel." Consistency between the calibration factor and EU name will prevent confusion and improve diagnostic accuracy.

Trigger Group

Many of the diagnostic techniques used in RTA rely on the ability to synchronize the event under investigation to some internal or external event. The trigger group setup parameter is used to define the specific event or variable that starts, or triggers, the data-acquisition sequence. Triggers, such as a once-per-revolution input from a tachometer, preselected time interval, or a variety of other sources, may be used to start the data-acquisition sequence.

When using an internal or external event to trigger data acquisition, the analyzer does not begin processing data until that event occurs. At that time, the analyzer acquires either a single block of data, or the requested number of samples (i.e., blocks) are collected.

The user can set parameters to perform sampling either (1) on the first trigger only or (2) on every trigger received. The user also can specify the characteristics of the trigger such as a signal coming from an external tachometer input or an analog signal coming from one of the channels. The user can control how soon data will be collected before or after the trigger occurs. The following information is required to set the trigger and data-acquisition characteristics: source, slope, threshold, and source-channel delay.

Source

The trigger source is selected using the source-combo box. There are four source options: free run, external, reference channel, and repetitive.

Free Run

In the free-run mode, data are acquired constantly at the maximum rate allowable given the number of tasks being processed by the program, which permits multitasking or running more than one application at a time. The number of applications directly affects the speed of data acquisition and display. For fast transient applications, the number of simultaneous applications should be kept to the minimum required to complete the process.

External

With external triggering, data are acquired time-relative to a TTL input signal on the dedicated trigger channel. The real-time analyzer has a dedicated channel for conditioned TTL tachometer input. This channel is in addition to the two to eight channels available for data acquisition. As an example, the Scientific Atlanta SC390 unit must have a TTL input to trigger data acquisition in the external trigger mode.

Reference or Internal Channel

With internal channel triggering, data are acquired relative to the input signal on the specified reference channel. When operating in two-channel, 100-kHz mode, the first channel must be the reference. All other applications can use any channel as the reference or trigger channel.

Repetitive

The repetitive check box triggering option determines if the data will be collected only on the first trigger or on each successive trigger. This option can be used for either the external or reference channel source selection. It cannot be enabled for the free-run option.

Slope

Slope defines the type of edge, either rising or falling, to be used for the trigger. Used in conjunction with the trigger threshold, the slope eliminates ambiguity in the specification for analog signal triggering.

With a rising-edge slope, increasing voltage of the signal at the specified threshold level serves as the trigger. With a falling-edge slope, decreasing voltage of the signal at the specified threshold level serves as the trigger.

Threshold

The threshold is used to set the trigger point for analog signal triggering. It is usually specified in terms of a percentage of the channel's full-scale RMS value. Any signal with the appropriate slope exceeding the threshold voltage will act as a trigger. The threshold is usually an integer percentage in the range of ± 99%, adjustable in increments of 1%.

Source-Channel Delay

The source-channel delay text box option is used to set the delay of the sample count between the trigger and the start of data collection. This delay setting applies to all channels and is an integer not to exceed the extended recorder memory size option.

A positive delay causes post-triggering, whereby data acquisition is delayed for some period after the trigger event. A negative delay results in pre-triggering or acquisition at some selected interval before the anticipated trigger event.

Chapter 26

ANALYSIS SETUP

In addition to the data-acquisition parameters discussed earlier, the analyst also must establish the parameters that will be used to analyze the data. Care must be taken to ensure compatibility between the acquisition and analysis setups.

Analysis mode can be used in conjunction with acquisition mode to view real-time data during the data-acquisition sequence. In this way, the user can monitor the vibration characteristics of the machine-train in real time. In addition, the user can verify the validity of data as they acquired it.

As with the acquisition mode, the RTA program requests specific inputs to define the user-selected analysis parameters used to condition and display the data. The menu-driven template requires user inputs for the basic setup, as well as the display setup.

BASIC SETUP

This section describes the basic setup required for a microprocessor-based, real-time analyzer. It must be completed each time a data set is evaluated or any time the active parameters change. Setup includes the following parameters: active channels, reference channel(s), block size, overlap, process weighting, and average group.

Active Channels

The active channels check boxes are used to select which channels will be used for data collection, conditioning, and display. Those channels not designated as active will be ignored, thus freeing memory for use by active channels. There must be at least one active channel at any given time, but the number is limited only by the analyzer hardware configuration. There are typically up to eight active channels.

The analyzer automatically resets the maximum frequency (F_{MAX}) to 100 kHz (two-channel operation) or 40 kHz (three- to eight-channel operation) when the user does not specify a F_{MAX} below these values.

Reference Channels

The reference channel selector only appears when cross spectrum is chosen as the analysis method. The reference channel option is used to select which of the active channels will be used as the reference channel for multichannel analyses, such as transfer functions and cross-products. Note that only a channel already designated as active can be used as a reference channel.

Block Size

The analyzer divides the continuous stream of data it collects into blocks to facilitate processing. The block size selection determines (1) lines of resolution when the magnitude spectrum (FFT) option is selected or (2) sample size when the time traces or compressed time options are selected. The block size options include the following:

512 samples	200 lines
1024 samples	400 lines
2048 samples	800 lines
4096 samples	1600 lines

Overlap

The overlap parameter is used to determine the percentage of overlap that will be used to speed up the data-acquisition and -processing time. As with the conventional single-channel, data-acquisition system, overlap averaging truncates the acquisition of one block of data and starts the acquisition of the next. Most analyzers permit the following overlap percentage selections: 0, 25, 50, 75, and 90.

Overlap averaging reduces the accuracy of acquired data and must be used with caution. Except in those cases where fast transients or other unique machine-train characteristics require artificial means of reducing the data acquisition and processing time, overlap averaging should be avoided.

A logical approach is to reduce or eliminate averaging altogether. Acquiring a single block or sample of data reduces the data-acquisition time to its minimum. In most cases, this time interval is less than the best time required to acquire two or more blocks using the maximum overlap sampling techniques. Eliminating averaging generally provides more accurate data.

No Overlap

When zero or no overlap is selected, the real-time analyzer always acquires complete blocks of new data. The data trace update rate is the same as the block processing rate. This rate is governed by the physical requirements that are internally driven by the frequency range of the requested data.

25 Percent

When 25% overlap is selected, the analyzer truncates data acquisition when 75% of each block of new data is acquired. The last 25% of the previous sample is added to the new sample before processing is begun. As a result, data accuracy may be reduced by as much as 25% for each data set.

50 Percent

When 50% overlap is selected, the analyzer adds the last 50% of the previous block to a new 50%, or half-block, of data for each sample. When the required number of samples is acquired and processed, the analyzer averages the data set. Accuracy may be reduced by 50%.

75 Percent

When 75% overlap is selected, each block of data is limited to 25% new data and the last 75% of the previous block. At 75% overlap, there is a potential for distortion of data.

90 Percent

When 90% overlap is selected, each block contains 10% new data and the last 90% of the previous block. Accuracy of average data using 90% overlap is highly questionable because each block used to create the average contains only 10% actual data and 90% of one or more blocks that was extrapolated from a 10% sample.

Process Weighting

The process weighting option controls the type of weighting function to be applied in performing FFT on blocks of time data. Weighting typically is used to reduce sources of analytical error.

The length of the signal represented by the block of time data may not be an integral multiple of the signal's period. Because the FFT is meaningful only on periodic signals, this means that the resultant FFT may not accurately represent the actual frequency of the machine-train. In this case, weighting can be used to modify the time block to artificially produce periodicity so that the resultant spectrum is much closer to the actual signature generated by the machine-train. Weighting options include the following: rectangular, Hanning, flat-top, and response.

Rectangular Weighting Option

The rectangular option does not weight the input signal. The values displayed by the real-time analyzer are identical to the raw signal generated by the transducer. With this option, raw time-waveform data are converted directly into a frequency-domain signature through FFT.

Hanning Weighting Option

The Hanning weighting option provides best capture of the individual frequency components that make up a signature. However, it may distort the actual amplitude of the frequency components. This weighting factor is normally used for magnitude spectra or normal FFT analysis.

Flat-Top Weighting Option

Flat-top weighting provides the best representation of the actual amplitude of each frequency component within a FFT. However, it may distort the actual location (i.e., frequency) of each and, therefore, is not normally used for magnitude spectra analysis.

Flat-top weighting is useful with cascade, or waterfall, analysis. Its conversion methodology modifies the profile of each frequency component so the true amplitude is displayed. Even though the actual location (i.e., frequency) of each component may be slightly out of position, the profile is more visible when closely packed in a waterfall, or cascade, display. The primary advantage of flat-top weighting is that it provides the ability to see the dominant frequencies in a waterfall format.

Response Weighting Option

Response is a special weighting factor that should not be used for vibration analysis. This method of weighting acquired data is limited to those applications where the impact response characteristics of various materials or structures are the subject of analysis. This methodology provides the ability to dampen signals caused by postimpact ringing of the object being tested.

Average Group

As with other types of vibration analyzers, most real-time analyzers provide the ability to average multiple blocks of data to derive a display of time-waveform or frequency-domain signatures that best represent the vibration generated by a machine. The averager is used to set the analyzer's ensemble averaging parameters. Its setup template requires user selection of the following options: domain, method, stop criterion, and stop time/count.

Domain Option

The domain information box identifies which averaging method is activated. The domain is set by the analysis mode and the averager partially determines the type of analysis and display that can be produced. (See Data Sources in the Display dialog box in the Help menu for more information.)

Time Domain Option

With the time domain setting, the averager is configured for averaging time-domain data. This setting must be used for synchronous spectrum displays and is selected by enabling the sync spectrum analysis mode. Time domain must also be used for compressed and normal time trace displays.

Spectrum Option

The averager is configured for power spectrum domain data by enabling the spectrum analysis mode. When this option is selected, the analyzer automatically captures the selected blocks of data, averages their power content, and displays a frequency-domain signature.

Cross-Properties Option

The averager is configured for power spectrum and cross-properties (i.e., transfer functions, cross-power spectrum, etc.) data by enabling the cross spectrum analysis mode.

None Option

When rapid transients, impacts, and a variety of other events are suspected, the use of averaging can distort data and prevent proper analysis. As a result, most real-time analyzers have the ability to acquire single blocks of data. By selecting the None option, the averager is disabled and the analyzer does not perform any averaging.

Method

Ensemble averaging is analogous to traditional number averaging. It combines a sequence of traces to produce a single trace that is an "average" of the sequence. It is much like the averaging used in single-channel meters. Averaging reduces or eliminates spurious signals or distortion caused by outside influences or nonrecurring events. The use of averaging techniques in RTA has the same advantages and disadvantages as in single-channel systems.

Averaging can be used to smooth out variations in the signal spectrum to determine typical or average signal behavior. In a similar manner, time-domain averaging also can help eliminate the effects of random signal noise. During time-domain averaging, the meaningful signal component tends to reinforce itself while the noise component tends to cancel itself.

The disadvantage of averaging is that transients that may be useful in isolating a problem are likely to be eliminated. The real-time analyzer, like the single-channel analyzer, rejects or averages out nonrepetitive data components, even those that may be key to the diagnostic being performed. Therefore, the use of averaging techniques for RTA should be limited to those applications where averaged data permits separation of extraneous noise or influences from the data set. The real-time analyzer supports three types of averaging: linear, exponential, and peak hold.

Linear Averaging

With linear averaging, each trace contributes equally. The advantage is that it is fast to compute, but is suitable only for shorter trace sequences. It is not effective for longer sequences because the average tends to stabilize due to the analyzer's finite resolution.

Exponential Averaging

With exponential averaging, traces do not contribute equally to the average. New traces are weighted more heavily than old ones. The advantage is that it can be used indefinitely. Because the average does not converge to some value and then remain relatively constant, this averaging method is not limited by sequence size or duration. The average dynamically responds to the influence of new traces and gradually ignores the effects of older traces.

If averaging is required, exponential averaging should be used for long-duration transients (e.g., complete coil run in a cold-reduction mill). Linear averaging should be used for short-duration transients (e.g., spindle acceleration).

Peak Hold Averaging

Peak hold is not a true averaging method. Instead, the "average" produced by this method is the highest value recorded in a select number of data blocks. Peak hold is useful for maintaining a record of the highest values attained at each point throughout the sequence of traces. The ability to capture and display these maximum values is a useful diagnostic tool that can be used to improve the accuracy of analysis.

Stop Criterion

The stop criterion combo box is used to select whether the averager stops on the basis of elapsed time (i.e., time), number of samples collected (i.e., count), or if a single average is performed (i.e., single).

Time

With a "time" stop criterion, averaging stops after the specified number of seconds have elapsed. This is the preferred criterion where the interval or duration of the acquisition is known (e.g., taped data).

Count

With a "count" stop criterion, averaging stops after the selected number of samples has been collected. This is the preferred method when the time data are continuous or when the measurement size is defined.

Single

With the "single" stop criterion, averaging is performed manually on a trace-by-trace basis. This is the preferred method for interactively selecting ensembles based on more complex criteria. Single averaging must be used with the single trigger acquisition option.

Stop Time/Count

The stop time/count scroll bar is used to set the number of seconds or ensembles that will terminate the acquisition and averaging of data. The range that can be selected for time is 1 to 86,400 sec (24 hr). The user can scroll upward or downward in increments of 1 sec.

The range that can be selected for count is 1 to 10,000 ensembles. The up-arrow and down-arrow icons permit scrolling in either direction in increments of 1 ensemble.

DISPLAY SETUP

The final setup function required to use a microprocessor-based, real-time analyzer for analysis is the display format. The display group is used to set up the type and data source for the analyzer displays.

The Windows-based operating system used by most of these instruments permits multiple simultaneous displays in any combination of formats. However, the number of active displays has a direct effect on the speed of both data acquisition and display update. Therefore, caution should be used to limit the number of active displays used in order to limit the reduction of real-time functionality.

The built-in screen on the real-time analyzer is somewhat limited when multiple displays are activated. Where the full-display functionality of the analyzer is needed for analysis, it is better to use an external monitor instead of the built-in screen.

Display Type

The display type option allows the user to select which type of trace to display from data currently in the analyzer's memory. A pop-up menu, which is activated by clicking on the type field, provides the display options that are available. Only valid types are listed and these vary depending on averager domain and source.

Averager Domain/Store Contents

Averager domain is specified in the average group selected by the user in the analysis setup. Store contents is defined by the current status of store memory.

Source

The source button group is used to determine if the trace data are live, come from the averager, or come from the storage memory. If a particular source is invalid for the selected display type, the corresponding button will be grayed out and will not permit the user to select it.

X-Axis Display

The analyzer's X-axis display controls the annotation and scaling of the display's X-axis. However, it is somewhat limited in the options that are available to the user. Unlike the single-channel system, the real-time analyzer does not permit orders (i.e., order tracking) as an optional X-axis value. However, certain spectrum analyzers (e.g., Scientific Atlanta SD390) provide an optional signal ratio analysis (SRA) module that permits order tracking. Because the ability to display spectral FFT data in orders or multiples of running speed is a useful diagnostic tool, users of the real-time analyzer may elect to add the SRA option to their existing analyzers.

With the exception of time-domain traces, orbits, Nyquist, and the correlation functions, the real-time analyzer permits X-axis displays in either Hertz (Hz) or cycles per minute (cpm). The user may select either of these units or change the units at any time to aid diagnostics.

Time-domain displays, both normal and compressed time, must be displayed in seconds. Orbit displays are limited to raw voltage (V) or engineering units (EU). Nyquist functions must be displayed in either volts/volts or engineering units/engineering units. Correlations, both auto and cross, are limited to seconds.

The scaling button group determines the axis scale. The user must select either linear or logarithmic (base 10). If one of the scaling types is invalid for the selected display type, it defaults to the valid scaling type and the control corresponding to the invalid type is grayed to make it unavailable as a selection.

Most domestic users will elect to use the linear scale. Single-channel systems normally default to the linear scale, so users tend to be more comfortable with this type of display.

Y-Axis Display

The Y-axis display options permit the user to select the vertical, or amplitude, scale and display type. Help messages and users' manuals describe the authorized display types for each of the data types that can be selected.

While the analyzer is somewhat more flexible in the options available for the Y-axis display, this is not universal. The user should review the options provided in the documentation to ensure proper selection. In addition, care must be taken to ensure that the data acquisition and analysis setups agree with the units and scaling selected for the display. The analyzer may display data in formats that are not technically correct when there is a conflict between the various user-selected setup options.

Units

The units option allows the user to select the units for the display's Y-axis. Selection depends on the display type and the axis scaling. The analyzer will default to normal units for each data type unless another option is selected. Note that the analyzer will not convert acquired units into other terms unless the unit types option is initialized. If the acquisition setup established that the data conversion was 100 mV/g, the analyzer will retain that conversion factor and display data in either g's or EUs.

Unit Type

The unit type option is available only for magnitude spectrum (FFTs) and provides the ability to automatically convert acquired data into other unit types. For example, data acquired as acceleration units (i.e., g's) can be integrated into velocity units (i.e., in./sec) and double integrated into displacement units (i.e., mils). The user may select the appropriate conversion method from the options displayed in the pop-up window adjacent to the unit type display. These options are normal, integrate, and differentiate. With the normal option, no change is made to the acquired data. If the integrate option is selected, the acquired data are integrated, but the user must select the resultant units (i.e., g's to in./sec, g's to mm/sec). The differentiate option converts data to the desired terms, but the user must select the resultant units (i.e., in./sec to g's, mm/sec to g's).

A real-time analyzer is not as flexible with units as a single-channel system. While the single-channel system permits selection of RMS, peak, or peak-to-peak scaling factors for each unit type, the real-time analyzer is limited to its default values. Most of the unit types are limited to RMS values, which can result in confusion. Many of the vibration-severity charts use a combination of unit qualifiers (RMS, peak, etc.) and the limited display capabilities of the real-time analyzer may result in the inability to compare displayed data to vibration severity charts.

Scaling

While the user may select either linear or logarithmic scaling, in most cases, the linear selection should be used for clarity. In addition, the user should always select the same vertical and horizontal scale. Mixing linear and log scales may distort the vibration profiles and cause misdiagnosis of the machine's condition.

Chapter 27

TRANSIENT (WATERFALL) ANALYSIS

RTA is ideally suited to both data capture and analysis of the short-duration events, or transients, that directly or indirectly affect machine reliability. Transient analysis (TA) is a common application of RTA. The most popular method of TA, waterfall analysis, captures a series of closely timed signatures to evaluate the transient. A waterfall, or cascade, plot like the one in Figure 27.1 is simply a display of multiple frequency-domain vibration signatures. In RTA, each signature is acquired and displayed as the event happens.

Pseudo-waterfall analysis can be conducted with some general-purpose vibration instruments. However, the real-time analyzer's combination of parallel multichannel signals and fast data-processing time makes it an excellent diagnostic tool.

SETUP

Most real-time analyzers provide the ability to capture and display a variety of waterfall or multiple-spectra plots that facilitate transient diagnostics. The setup procedures for this type of analysis may vary, depending on the type of instrument used. However, the following information is usually required by all: input group, load-control group, load method, and number of records.

Input Group

The input group setting defines which of the recorded data channels will be used for analysis. Typically, a real-time analyzer has between two and eight channels that can be used to record, condition, and display waterfall data.

Active Channels

Active channels are used for waterfall data collection and analysis. Those channels not designated as active are ignored. It is important to use only those channels actually

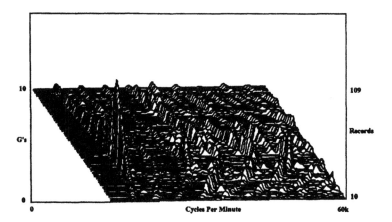

Figure 27.1 Multiple frequency-domain vibration signatures.

required for analysis, because the number of active channels directly affects the speed of data acquisition and processing, as well as the extended memory required.

Reference Channels

A reference channel is needed for multichannel analyses, such as transfer functions. Only a channel already designated as an analysis reference channel can be used as the reference channel in waterfall analysis. To add additional reference channels in the waterfall mode, reset the acquisition and analysis setup templates to include the additional channels.

Function Group

The function group setting allows selection of the waterfall analysis function that will be loaded into waterfall memory.

Mode

Mode determines which types of analysis functions are available to be waterfalls. This parameter cannot be changed on this template.

Function

Function is a pull-down box displaying the choices of functions that can be loaded into waterfall memory for the selected mode. Only one function can be selected.

Source

The source setting defines the source of data to be included in the waterfall. One option is live or real-time data, which are either directly from the machine-train or from prerecorded (taped) data. These data are single block and weighted using the selections of prior setup steps. The second option is averager data. If selected, this option provides averaged data for the waterfall display.

In most waterfall applications, real-time data are preferred. Because the method of analysis is used predominantly for transient analysis, real-time, single-block data provides the best diagnostic capability.

Load Control

Load control options determine how data are transferred from the source-data area to the waterfall memory. The choices reflect the selected analyzer mode, data, source, and acquisition choices that have already been established as part of the previous setup functions. Options also include continuous, average recycle, delta time, percent amplitude, and delta rpm.

Continuous

Continuous control has two different functions based on whether the data source is live or averaged. Live data are transformed into the desired function and transferred to the waterfall memory as fast as the analyzer can collect and process the data. The rate of processing depends on the function, selected number of channels, and number of active displays. Data are always block processed, but with very small overlap. Averaged (AVG) data defaults to the averager control parameters (e.g., overlap, weighting, etc.). When one block of data is processed according to the averager overlap factor, the desired function is transferred to waterfall memory.

For most transient-capture applications, the averaged-data option is used for the waterfall. The default values for live data are used to provide continuous data for the waterfall memory and display update.

Average Recycle

Average recycle load control is only available when the data source is from the averager. Each time the averager target count reaches the specified number, the data in the averager are copied to the waterfall memory. Each copy produces one record for each selected waterfall channel. The averager is cleared and the process begins again until the requested number of records has been copied to the waterfall memory. Average recycle forces the averager to use the linear average and count stop method. Because a linear average is used, this method is not suitable for long-duration events and should not be used except in fast-transient capture events.

Delta Time

When delta T is selected, the time increment field appears to allow the user to enter the desired time increment (i.e., seconds) in the appropriate field. When the selected time increment elapses, data in the source memory are processed and transferred to the waterfall memory. The granularity of the time increment is 10 msec.

Percent Amplitude

When the amplitude load setting is selected, an amplitude value in percent of full scale can be entered. This determines when the data from the selected source are transformed and transferred to the waterfall memory. When the amplitude (must be RMS value) of the first selected waterfall channel from the selected waterfall source exceeds the indicated percentage of full-scale amplitude, data from all channels are processed and copied to the waterfall memory.

This option can be used to trigger FFT generation at selected energy levels. For example, this technique could be used to capture repetitive occurrences of resonance or other abnormally high vibration levels without processing all signatures that contain normal energy levels.

Delta RPM

The delta RPM option permits selection of specific rpm intervals where the data in the selected source are processed and transferred to the waterfall memory. This method is typically used to perform runup and coastdown transient analysis. For example, the user could elect to capture a FFT for each X rpm increase during runup and each X rpm decrease during coastdown.

Load Method

The load method (i.e., continuous or "stop when full") determines if waterfall memory will be loaded continuously or if loading will stop when waterfall memory is full. When loading is continuous, data wraps when waterfall memory is full. When this happens, the first FFT is dropped from the display and a new signature is added. This process continues until the user manually stops the process by pressing the "stop waterfall load" or hold button.

Number of Records

This parameter specifies the number of records per channel that will be collected before loading stops when using the "stop when full" load method, or before data wraps when using the continuous load method.

ANALYSIS

The diagnostic logic used to perform waterfall analysis is identical to that used in frequency-domain analysis. The only true difference is in the quality and timing of the data that are used.

In traditional frequency-domain analysis, data are acquired periodically over long intervals (i.e., weeks or months) of time. A series of signatures is then displayed so that a comparative analysis can be made.

RTA uses data that are acquired as the event happens. As a result, there is typically much more data, but its total time span is relatively short. For example, a typical transient through a critical speed, or resonant, zone may last less than 15 sec. However, the real-time data may contain 50 or more individual signatures.

Chapter 28

SYNCHRONOUS TIME AVERAGING

The signal to be measured and analyzed can sometimes be corrupted by unwanted contributions from noise, line hum, or other machines running nearby. When this occurs, spectrum averaging can be used to smooth out the total signal, but the technique does not necessarily help with analyzing a particular machine.

Other techniques are required to extract only the signal harmonics related to the desired rotating element from a composite measured signal.

SEPARATING SIGNAL FROM NOISE

Several techniques are available for cleaning up a signal surrounded by noise. With most techniques, some type of spectrum averaging is used. However, for the most part, the enhancement provided by standard averaging is limited. If the level of the periodic component is below that of the surrounding noise, increasing the number of spectrum averages will not reveal the buried signal. It will only make the adjacent noise smoother.

Figure 28.1 illustrates a periodic signal that has been buried in noise. The display on the left is the unaltered waveform. The periodic signal also was available as a reference waveform, but was not used in this first measurement. The display on the right is that of a standard averaged spectrum. In this case, 1000 averages were used and the total signal power was averaged. In the standard averaged display, only the fundamental spectrum component at 350 Hz is visible. From this display, it is not clear if there are any harmonics or their distribution.

With synchronous time averaging, blocks of time records are triggered from the desired reference signal and averaging takes place in the time domain. Waveforms

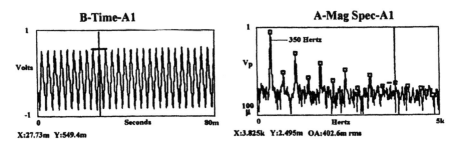

Figure 28.1 Unaltered time waveform (left) and standard averaged spectrum (right) after 1000 averages.

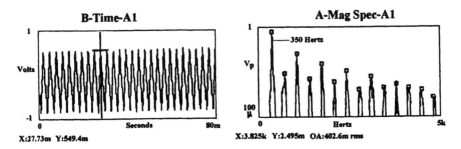

Figure 28.2 Sync averaged time waveform (left) and sync spectrum (right) after 10,000 averages.

that are synchronous with the reference tend to be reinforced frame after frame, whereas those that are random or synchronous at a different rate are not reinforced and average to zero. After sufficient averaging in the time domain, a single FFT is performed and the result is referred to as a sync spectrum.

With this method, the signal-to-noise ratio (S/N) improves proportional to the square root of the number of averages taken. Theoretically, all noise components could be eliminated by averaging for a long period of time. The result of synchronous spectrum averaging should be only a periodic spectrum or nothing at all. Because the reference signal is available for the waveform shown in Figure 28.1, it can be used as an external trigger input to the real-time analyzer and a Synchronous Average performed (see Figure 28.2).

Two important factors should be remembered when performing a synchronous time average:

    1. A reference signal is crucial and must be mechanically linked to the physical shaft or other rotating element to be isolated. An external signal generator set, for example, at approximately the same frequency as the desired rotating element turning speed may appear to momentarily offer some sig-

nal cleanup. However, since it is not phase locked with the machine, averaging enhancement soon ceases and no long-term improvement results.

2. Averaging is done in the time domain and must be followed by a single FFT. This is important because the FFT typically performs a power average, including all coefficients that are present. However, with synchronous averaging, we care only about terms that are phase locked with the reference signal. Thus, we need to average in the time domain by creating blocks of data that always begin (or end) when the tachometer triggers. In this way, nonsynchronous, or random, events average to zero due to bipolar cancellation. In nonsynchronous or standard spectrum averaging, all components (both synchronous and other periodic and random data) are averaged. This smooths the spectrum, but does not enhance the S/N.

APPLICATIONS

There are many applications of synchronous time averaging, but the more common uses include (1) obtaining multiple-order reference, (2) eliminating beating responses, and (3) recovering signals below the noise level.

Multiple-Order Reference

In some cases, the primary signal of interest may not be at the fundamental shaft-speed frequency, but rather at some multiple of that frequency. For example, the analyst may want to focus on the condition of a rotating element by removing all vibration except for specific passing frequencies, such as the number of blades on a turbine wheel, the number of impellers in a pump, gear ratio, etc.

If a multiple tachometer signal of the desired rate is available, it can be used as the reference trigger input. By using multiples of the tachometer input, only the trigger multiple and its harmonics will be preserved. An example of this is seen in Figure 28.3 below, where a 3× running speed signal was used as the trigger input to the time synchronous averaging process. With this trigger, the time waveform and resulting

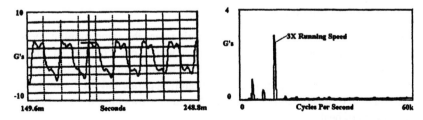

Figure 28.3 Sync averaged time waveform (left) and sync spectrum (right) using 3× as trigger.

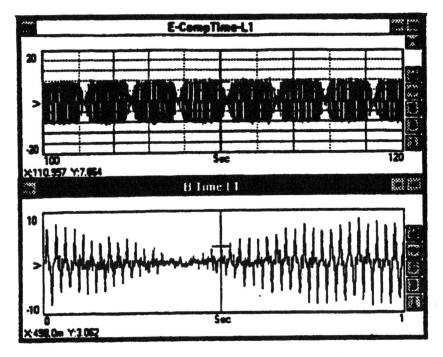

Figure 28.4 Long-term (top) and expanded (bottom) time recording of beat.

spectrum will contain only the third order of running speed and each of its harmonics (3×, 6×, 9×, etc.).

If the contribution of a particular component (e.g., 35-tooth gear) is of interest, it also is feasible to multiply a known tachometer signal by a phase-locked ratio to recover just that signal from a composite waveform. In this case, the higher harmonics may not be of interest and, if they are, there may be a conflict between spectrum resolution and waveform stability. Remember that time synchronous averaging depends on the stability of the systems generating the multiple signals, especially on the stability of the recovered waveform. The more this waveform wanders in frequency or speed, the more difficult the synchronous averaging process will be.

Eliminating Beating Responses

In some cases, two machines or elements operating near the same running speed can create a phenomenon referred to as "beating," which can confuse an otherwise straightforward measurement. This is another case where having the tachometer reference for the specific rolling element can accurately separate out the desired component. An example of this is seen in Figure 28.4, which shows long-term and expanded time recordings of beat.

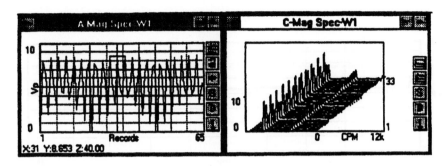

Figure 28.5 A 2400-cpm, 40-Hz track (left) and 3-D waterfall (right) over a 1-min interval.

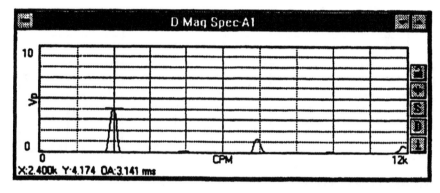

Figure 28.6 Sync spectrum using 2400-cpm reference, 200 averages; beating component at 2376 cpm.

Figure 28.5 illustrates a 3-D waterfall and 2400-cpm, 40-Hz track over a 1-min interval. Note that beating causes the spectrum level of the 2400-cpm, 40-Hz component to vary from nearly 0 to more than 8.6 V.

However, as shown in Figure 28.6, the actual level of the 2400-cpm component, as derived through sync averaging with respect to the actual 2400-cpm reference, is 4.174 V peak.

Recovering Signals Below the Noise Level

A benefit of synchronous time averaging is that no theoretical limit exists for the amount of signal "cleanup" that can be performed. It is one of the few signal-processing techniques available where the S/N enhancement is proportional to the square root of the number of time-averaged blocks processed. Thus, the longer the averaging

D-Mag Spec-A1

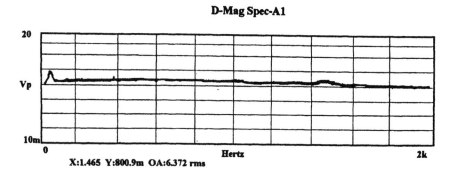

X:1.465 Y:800.9m OA:6.372 rms

Figure 28.7 Standard free-run spectrum analysis performed with 1000 spectrum averages.

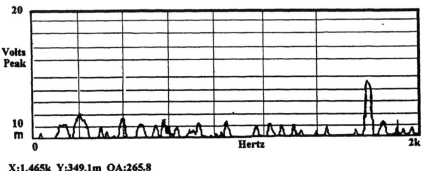

X:1.465k Y:349.1m OA:265.8

Figure 28.8 Sync averaged spectrum of same signal shown in Figure 28.7.

takes place, theoretically, the cleaner the signal gets until the only signal left is the trigger, or synchronizing frequency, and its harmonics.

The spectrum shown in Figure 28.7 is the result of 1000 averages of a free-running signal input. The only apparent signal is a peak at 40 Hz. However, it is suspected that there might be a signal contributing at approximately 37 times the nominal 40-Hz component. To verify this, a synchronous time average was performed with a reference signal of 1466 Hz.

The time synchronous spectrum shown in Figure 28.8 was performed with 3400 time averages followed by a single FFT. This technique makes it apparent that a clear signal exists at 1466 Hz. Also note that the amplitude of this 1466 Hz component is less than half that of the amplitude of the 1466 Hz component in Figure 28.7. This indicates that the desired signal is at least 6 dB below the level of the surrounding noise in the original broadband spectrum.

Chapter 29

ZOOM ANALYSIS

Zoom analysis provides the means to separate quickly machine-train components, such as gear sets, from a complex vibration signature. The technique lets the user select a specific range of vibration frequencies, which the real-time analyzer converts to a high-resolution, narrowband signature. This capability is unique to real-time analyzers and is not available in general-purpose, single-channel vibration analyzers.

Real-time zoom analysis can be performed with no data gaps up to a range of 10 kHz with most microprocessor-based, real-time analyzers. However, the center frequency plus one-half of the selected frequency span cannot exceed 10 kHz.

Above this range, pseudo-real-time processing occurs, which means that data required to perform the zoom transform are acquired until the extended recorded memory of the analyzer is full. When this occurs, the acquired data are processed before additional data are gathered. However, this may result in data gaps that can adversely affect the accuracy of the zoomed spectra. The gaps will be proportional to the time required to perform the zoom transform for each channel, which in some cases can be between 5 and 10 sec.

When using the zoom mode, the extended recorder memory should be set to the maximum available to obtain the best zoom accuracy and resolution. Reducing the number of active channels and lines of resolution also increases the speed and minimizes the data gaps.

FREQUENCY SPAN

The frequency span parameter allows the user to select the frequency span for spectrum (FFT-based) and octave (digital filter-based) acquisition and analysis. For spectrum analysis, the frequency span can be set to any frequency notch from 1 Hz to 100 kHz (usually limited to two-channel operation only) or 1 Hz to 40 kHz (for three- to eight-channel operation). In real-time zoom mode, the frequency span can be set to any frequency notch from 5 Hz to 10 kHz, as long as the new frequency span is in the range of zoom capabilities.

CENTER FREQUENCY

The center frequency setting is used to set the center frequency for zoom mode operation. The center frequency can be set to any value in the range up to 100 kHz – (Frequency span/2) for two-channel operation or 40 kHz – (Frequency span/2) for three- to eight-channel operation.

Chapter 30

TORSIONAL ANALYSIS

Torsional vibration is not a simple parameter to analyze because transducer requirements are stringent and shaft access may be limited. In addition, there is a peculiar mystique engulfing torsional vibration. This chapter attempts to clarify the process of its analysis through experimental examples and descriptions of the basic fundamentals of torsional motion and how it can be interpreted.

WHAT IS TORSIONAL VIBRATION?

Torsional vibration of a rotating element is the rapid fluctuation of angular shaft velocity. As a machine changes speed, torque is applied to the shaft in one direction or the other. A machine often increases or decreases speed over some period: weeks, days, or seconds. However, when the rotational speed of the machine fluctuates during one rotation of the shaft, it is considered torsional vibration. Because this type of vibration involves angular motion, the basic units are either radians or degrees.

Figure 30.1 shows the end view of a shaft in a bearing with a position marker, called a key-phasor. An angular reference grid that is marked in 10-degree divisions surrounds the shaft. In this example, an operating speed of 0.1667 rpm is assumed. This is equivalent to a rotational rate of one revolution in 6 min, or 1 degree/sec (true only if there is no torsional vibration). If the shaft turns at a constant rate of 1 degree/sec, then the angular velocity is constant. No torsional vibration can be present under this condition.

As an example of a shaft experiencing sinusoidal angular velocity changes, assume a rotating shaft increases to a maximum turning rate of 1.06 degree/sec during the first 10 sec of rotation. Also assume that it slows to a minimum rate of 0.94 degree/sec during the next 10-sec period. Under this condition, this shaft experiences the torsional vibration, frequency, and amplitude shown in Figure 30.2.

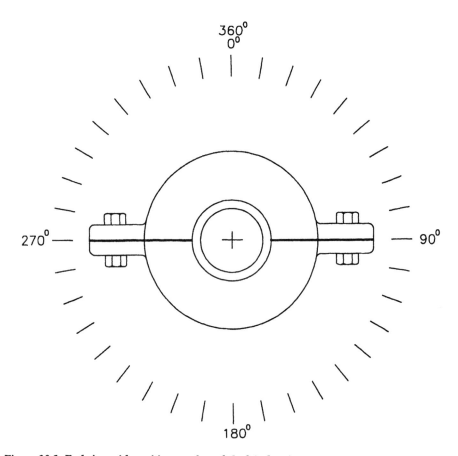

Figure 30.1 End view with position marker of shaft in bearing.

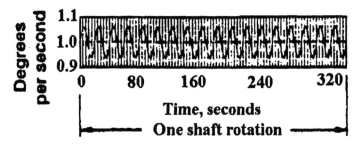

Figure 30.2 Torsional vibration graph.

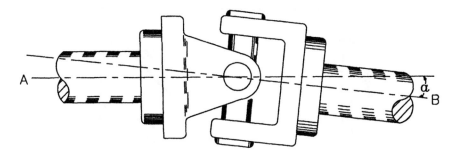

Figure 30.3 Hooke's joint.

In this example, both shafts complete one rotation in 1 sec. If we could look at an rpm readout for each shaft, we would see they are both turning at the same speed. The first shaft turns at a constant rate of 1 degree/sec. The second shaft turns at an average rate of 1 degree/sec. The torsional vibration waveform (see Figure 30.2) goes through one complete cycle every 20 sec. This is 18 cycles per revolution, which corresponds to a frequency of 0.05 Hz or 3 rpm.

STANDARDS

The basic standard for analyzing torsional vibration is the universal joint (see Figure 30.3), which also is known as Hooke's joint. This device produces a predictable value of angular velocity, rpm_B, for output shaft, B, referenced to a constant speed, rpm_A, of input shaft, A, and the angle, α, between the centerlines of the two shafts. It may be difficult to comprehend, but at the instant of time portrayed in Figure 30.3, shaft B is turning at its maximum rate, which is faster than shaft A. After shaft A rotates 90 degrees, shaft B turns at its minimum rate, which is slower than shaft A. Two complete maximum/minimum cycles of shaft B occur every 360-degree rotation of shaft A. This means the torsional frequency is twice the rotating speed.

Figures 30.4 and 30.5 illustrate how the angular velocity vibration level of the output shaft changes with input shaft speed, rpm_A, and shaft angle, α. Notice the linear relationship between angular velocity vibration level and shaft speed in Figure 30.4. In Figure 30.5 the angular velocity vibration level increases exponentially as the shaft angle, α, increases at a linear rate. Angular velocity vibration levels are expressed in units of degree/sec, peak.

Now that we have discussed changing shaft velocity, we will look at the rate of the changes. The rate at which the shaft changes its angular velocity is the measurement of angular acceleration (not normally used to express levels of torsional vibration). Angular acceleration is harder to comprehend and produces very large numbers. The results shown in Figures 30.6 and 30.7 are obtained by differentiat-

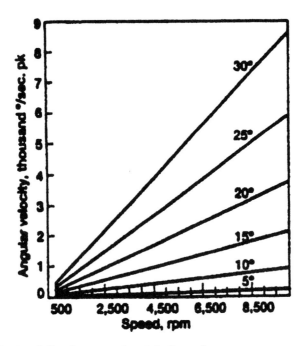

Figure 30.4 Torsional vibration versus input shaft speed.

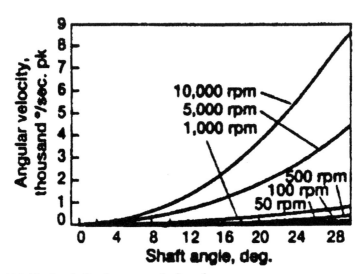

Figure 30.5 Torsional vibration versus shaft angle.

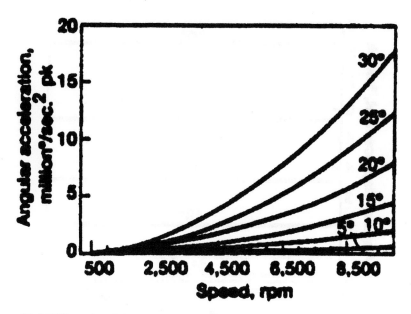

Figure 30.6 Differentiated data from Figure 30.4.

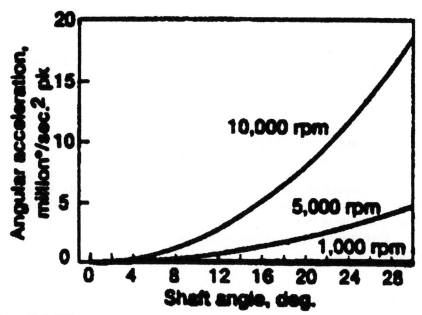

Figure 30.7 Differentiated data from Figure 30.5.

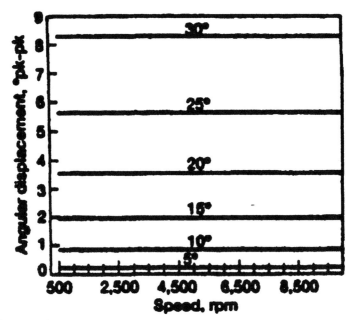

Figure 30.8 Angular displacement versus input shaft speed.

ing the data in Figures 30.4 and 30.5. Angular acceleration is displayed in units of degree per second squared, peak.

The most commonly used parameter for expressing torsional motion is angular displacement, whose units are degrees, peak-to-peak. There are several reasons to express torsional motion in terms of angular displacement:

- Has a small numerical value.
- Tends to remain constant during speed deviations.
- Is easier to visualize motion.

Figures 30.8 and 30.9 show the angular displacement values produced by the Hooke's joint relative to input shaft speed and U-joint angle. It is obvious from the data that the torsional vibration is completely independent of input shaft speed. The torsional vibration amplitude of induced angular displacement depends solely on the U-joint angle, α. Even at angles up to 30 degrees, the vibration level is always in single-digit quantities.

DETERMINING TORSIONAL MOTION

Determining the torsional response of shafts or other components requires a positive means of measuring or calculating the movement of two reference points, one on each

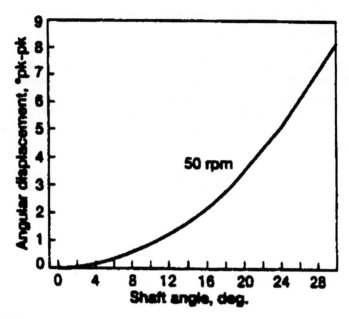

Figure 30.9 Angular displacement versus shaft angle.

end of the shaft. The following methods are used to measure the motion: optical encoder, gear teeth, charting and graphic art tape.

Optical Encoder

Torsional motion is a deviation in shaft speed during one revolution, which must be sensed instantaneously in order to be detected. This requires a signal of many pulses per revolution (ppr). The best way to measure abrupt changes in shaft velocity is with an optical shaft encoder, a device that consists of a spinning disk with very accurate markings.

The encoder normally connects directly to the end of a shaft. As the disk spins, the marks produce a pulse output each time they pass a photocell. The number of pulses per revolution depends on the application. The following are the main factors to consider with these devices:

- *Machine speed.* Optical encoders have frequency limitations and, for a specified pulse rate, there is a maximum turning speed.
- *Torsional component frequency.* This is a matter of resolution and, when selecting the proper encoder, the Nyquist sampling rate must be used. This problem must be dealt with any time digital sampling of analog data is undertaken. The Nyquist sampling theorem states: Data must be sampled at

a rate greater than two times the highest frequency content of the data being sampled. This keeps the data free of extraneous aliasing terms.

- *Torsional phase.* This measurement requires two signals, one from each end of the shaft. To measure static twist, signals having one pulse per revolution are sufficient. When measuring the phase of dynamic frequencies, a multiple pulse rate is required. The same sampling constraints for torsional component frequencies apply when making phase measurements.

Gear Teeth

The use of gear teeth is sometimes the only method that can be used to detect torsional motion. The advantages make this technique worth considering, but watch out for the pitfalls.

Charting and Graphic Art Tape

Graphic art tape is a photo tape with very accurate black and white bars running across it. It is available in many art supply, and some stationary stores. This tape can be used with an optical sensor to detect shaft speed changes.

The major advantage of the tape is that it can be wrapped around a shaft that lacks exposed ends. A serious drawback is the discontinuity point where the two ends of the tape join. This introduces a torsional component at the shaft speed frequency along with its associated harmonics.

Measured Versus Calculated Data

Measured data and calculated responses compare very well until the input shaft speed reaches 1560 rpm, at which point the curves begin to deviate. Why does this deviation occur? The Hooke's joint data must be acquired at low input shaft speeds to be reliable because, as speeds increase and angles get greater, the mass of the system cannot respond to the rapidly changing shaft velocities.

Measured Data

Figure 30.10 shows the data-acquisition arrangement from a Hooke's joint test stand in which a variable-speed dc motor drives input shaft A through a flexible coupling and optical encoder. Both the input and output shafts are 0.375 in. in diameter. Two bearings support shaft A and eight flywheels, each weighing 1 lb. The flywheels aid in maintaining a constant angular velocity for shaft A. The driving force for shaft B is supplied by shaft A through the U-joint. Two bearings support shaft B. The angle, α, between shafts A and B can be set to 0, 15, or 30 degrees. Shaft B drives a second optical encoder used to detect the torsional motion introduced by the U-joint.

Both optical encoders receive power from a torsional converter with two independent channels, each of which receives one of the 40-ppr optical encoder signals. The two

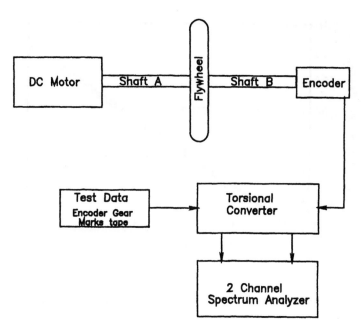

Figure 30.10 First analysis test stand.

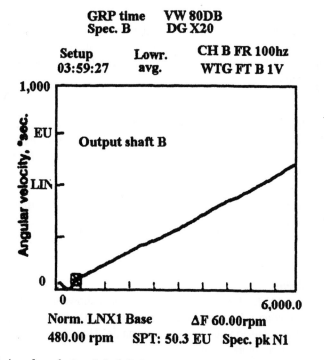

Figure 30.11 Angular velocity of shaft B data.

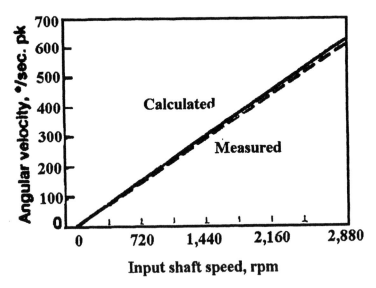

Figure 30.12 Calculated response versus measured data for 15-degree U-joint angle.

conditioned output signals are connected to a two-channel, real-time analyzer. The torsional converter converts the 40-ppr signal from the shaft A encoder to a 1-ppr signal to be used as a tachometer signal. The real-time analyzer also uses the 1-ppr signal to normalize changing frequency components to the input shaft speed for phase measurements.

Fifteen-Degree Angle
The following describes an example using the arrangement described in the preceding section with a U-joint angle of 15 degrees. In this example, the analyzer stores the data using peak-hold averaging. Figure 30.11 represents the angular velocity measurement of shaft B.

Figure 30.12 is generated by picking discrete data points every 120 rpm and superimposing them on the calculated response displayed from Figure 30.12. Allowing for analysis error due to filter width and weighting, plus the continuously changing speed in the test, the 5% deviation exhibited is acceptable.

Thirty-Degree Angle
Notice that the amplitude of the angular velocity signal shown in Figure 30.13, which represents a test having a U-joint angle of 30 degrees, has increased approximately four times the level obtained with the 15-degree angle. This agrees with the calculated response illustrated in Figure 30.14.

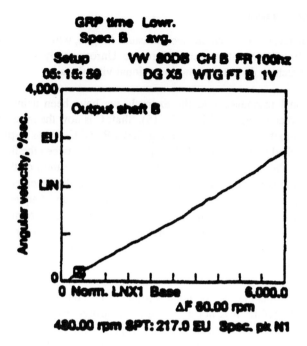

Figure 30.13 Increased angular velocity signal.

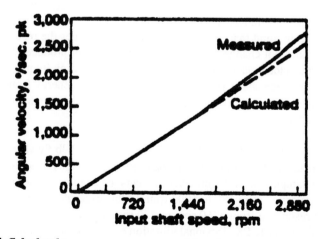

Figure 30.14 Calculated response versus measured data for 30-degree U-joint angle.

TORSIONAL RESONANCE

Like any resonance, torsional resonance can cause fatigue, which in turn leads to shaft cracking, coupling deterioration, gear failure, etc. Unfortunately, standard transducers mounted on a machine do not respond to torsional vibrations.

To build torsional resonance into the example we have been using, we must add another shaft and coupling (see Figure 30.15). Shaft C is now the mass of the resonant system and the flexible coupling connecting shafts B and C acts as a spring. An accelerometer is attached to the bearing that supports shaft B between the flexible coupling and the U-joint. This accelerometer detects any radial vibration of the bearing. The torsional vibration is monitored and displayed as angular displacement. Figure 30.16 shows two data traces. The top trace shows the amplitude and frequency of the bearing's radial motion. The bottom trace shows the amplitude and frequency of the torsional vibration. However, we are more interested in the frequencies present than their amplitudes. The accelerometer output shows four structural resonances not reflected in the torsional data.

The important point is that the torsional resonance at 5760 rpm is not detected by the accelerometer, therefore, it is not affecting the bearings. However, the amplitude of 9.56 degrees PK-PK is severe. Referring to Figure 30.8 you will find the torsional force applied by a U-joint at 15 degrees is 2 degrees PK-PK. This means that the torsional resonance amplifies the vibration level by a factor of 4.78 to 1.

MASS DAMPERS

The addition of mass dampers can help solve certain problems. However, adding them changes the torsional resonant frequency, which can cause problems if the resonant frequencies are moved closer to operating speeds or other torsional forcing functions. Keep in mind that, while the added mass changes the resonant frequency of shaft C, it does not change the torsional vibration level produced by the U-joint. Shaft C with the added mass only responds to lower excitation frequencies. The flexible coupling between shafts B and C absorbs most of the energy from the torsional input. This mass-damping principle is the technique used in torsional dampers and harmonic balancers.

To illustrate the concept of mass dampers, we will add a 1-lb flywheel to shaft C. A torsional displacement measurement can be used to gauge the impact of the use of the flywheel as a damping mass. Figure 30.17 reflects the angular displacement of shaft C as the test accelerates from 180 to 6000 rpm. In this example, the added mass moves the torsional resonance from 5760 rpm to 720 rpm, and increases its amplitude from 9.56 to 17.3 degrees PK-PK. Torsional vibration excitation forces from the U-joint are still just 2 degrees PK-PK.

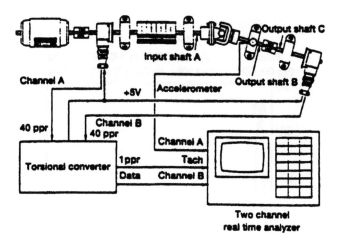

Figure 30.15 Second analysis test stand.

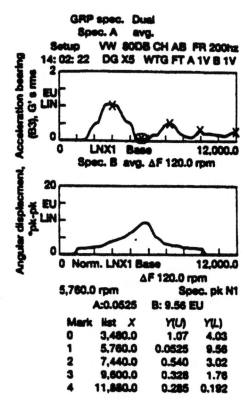

Figure 30.16 Bearing's radial motion (top) and torsional vibration (bottom).

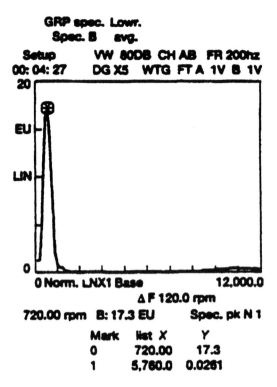

Figure 30.17 Angular displacement of shaft C with a one-pound mass damper.

Note that the frequency of these forces is twice the frequency of the shaft speed. If the machine operates above this frequency, the resonance will have no effect and the angular displacement observed in Figure 30.17 (9.56 degrees PK-PK at 5760 rpm) is reduced in amplitude to 0.0261 degrees PK-PK.

TORSIONAL PHASE

To sense the phase deviation from one end of a shaft to the other, it is necessary to measure the angular position of the shaft at each end. In Figure 30.18, an optical encoder is attached to both ends of a shaft.

The frequency of the encoders is 40 times the shaft speed and, for every 9 degrees the shaft turns, the encoder puts out one pulse (360(/40 ppr). The phase relationship between the two encoders at 855 rpm is shown in Figure 30.19. A phase reading of 0 degrees is ideal, but is next to impossible to achieve. It depends on the alignment of the two encoders. In the case of the test model, the reference phase measurement is 113 degrees when the two ends of the shaft are in phase. If a 9-degree twist develops

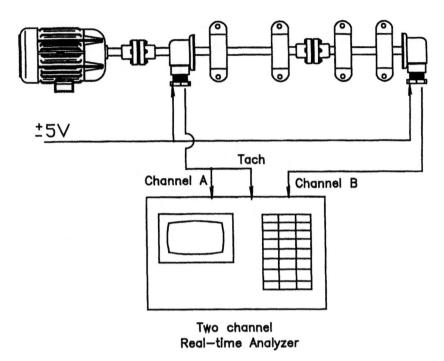

Two channel
Real—time Analyzer

Figure 30.18 Encoders at both shaft ends.

in the shaft, the encoder signals will rotate through 360 degrees and return to 133 degrees.

The count depends on the torsional frequency to be measured and the severity of the twist. Higher torsional frequencies require more pulses per revolution, while greater twists require fewer pulses per revolution. If the shaft under discussion in Figure 30.18 experiences more than 9 angular degrees of twist, an optical encoder of fewer pulses per revolution is needed.

LOOKING AT ENCODER OUTPUTS DIRECTLY

For this next example, we will run the test at 45 rpm (see Figure 30.19). The frequency of each encoder is 1800 rpm (encoder count, 40, times shaft speed, 45). Under ideal conditions, the phase offset should be constant. Figure 30.20 shows the phase between the two encoders over a 7-sec period. Notice that the phase shift at 45 rpm is the same as it was at 855 rpm and the phase varies between 130 and 137 degrees. This shows that there is a 133.5-degree offset between the two encoders and a 7-degree phase modulation present. However, the phase modulation is not as bad as it may appear. Based on the phase sensitivity, there are 0.175 angular degrees of error in our model due to shaft eccentricities, bearing drag, etc. (Angular phase = frequency phase shift/encoder count = $(137 - 130)/40 = 0.175$.)

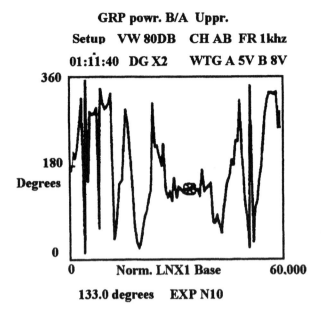

Figure 30.19 Phase relationship between encoders.

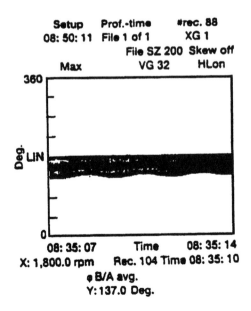

Figure 30.20 Phase over 7 sec.

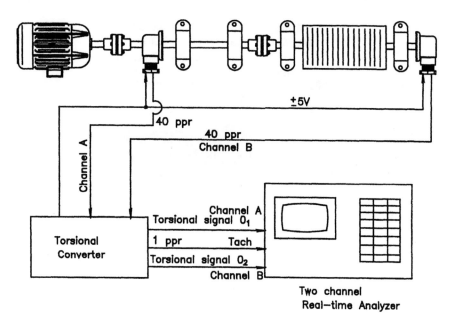

Figure 30.21 Third analysis test stand.

REASONS TO MEASURE

There are two reasons to measure torsional phase: static and dynamic twist. Although torsional phase can be used to pinpoint torsional resonance, this is not very important because it can be spotted just as easily using torsional amplitudes.

Static Twist

Shafts twist when under load. This is especially true during rapid speed changes, such as startup acceleration or braking on deceleration. As the shaft ages, the amount of twist increases and the shaft's ability to return to a neutral position decreases.

Take as an example the system shown in Figure 30.21. In this test, the flywheel on shaft B weighs 8 lb. The test will be run at idle condition before applying full power. During idle, shaft B neither leads nor lags shaft A and, by definition, the two shafts are in phase, although the phase angle may not read zero.

Adding power increases the speed. The phase reading between shaft A and B changes as the mass of shaft B resists the angular acceleration. A phase shift is generated by the twist that develops in the flexible coupling between the two shafts.

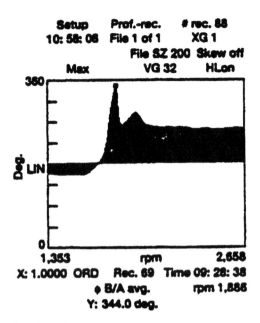

Figure 30.22 First-order phase data.

For data normalization, the signature ratio adapter built into the real-time analyzer tracks the various frequencies associated with the running speed of the test. This is necessary to compensate for any speed changes in the test.

The phase angle between the two encoders can be measured using the cross power spectrum and phase modes of the analyzer. Figure 30.22 is obtained by transferring the phase information to the analyzer waterfall and profiling the first order.

The data reflect a constant phase angle of 156 degrees during idle at 1353 rpm. Maximum torque occurs at 1886 rpm as the phase shifts to 344 degrees. The speed continues to increase to 2658 rpm, where the phase stabilizes at 256 degrees. There is a slight oscillation in the phase as the machine reaches full speed. This is caused by the accumulated torque stored in the flex coupling being released back into the shafts. Because damping is minimal, some overshoot occurs.

The final steady-state phase measurement differs from the original because of the added drag and friction caused by the new speed. The maximum amount of twist seen by the shaft as the torque is applied is calculated as follows: angular phase (AP) = frequency shift/encoder count = $(344 - 156)/40 = 4.7$ degrees of maximum twist.

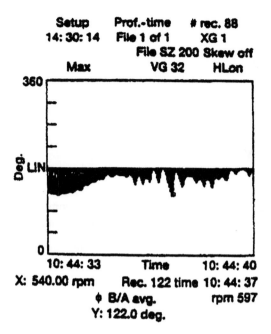

Figure 30.23 Results of varying load test.

Dynamic Twist

To measure a dynamic phase shift, a load is added between the bearings on shaft B and the test speed is adjusted to 540 rpm. The load is a manually activated mechanical brake operated at a random rate during the 7-sec data-acquisition period. Figure 30.23 shows the results of this test. Load variance causes varying amounts of twist to develop in the shaft at a rapid rate. This is a principal cause of fatigue in drive-train components. In this example, maximum deviation was 60 degrees, representing 1.5 degrees of twist. Note that all of the frequencies involved in shaft-twist deviations will be reflected in a torsional vibration spectrum.

GLOSSARY

Absolute fault limit	The maximum recommended level of overall vibration accepted in machinery.
Acceleration	The rate of change of velocity with respect to time.
Accelerometers	Use a piezoelectric crystal to convert mechanical energy into electrical signals.
Amplitude	The maximum absolute value attained by the disturbance of a wave or by any quantity.
Apex	The pitch angle is the sum of the pitch lines extended, which meet at a point.
Balance	All forces generated by or acting on the rotating element of a machine-train are in a state of equilibrium.
Ball spin frequency (BSF)	The spinning motion of the balls or rollers within a bearing.
Ball-pass inner-race (BPFI)	The ball/roller rotating speed relative to the inner race.
Ball-pass outer-race (BPFO)	The relative speed between the balls or rollers in a rolling-element bearing and the outer race.
Broadband	A band with a wide range of frequencies.
Broadband energy	Provides a gross approximation of machine's condition and its relative rate of degradation.
Broadband trending	Vibration analysis technique that plots the change in the overall or broadband vibration of a machine-train.

Centrifugal pump	A machine for moving a liquid, such as water, by accelerating it radially outward in an impeller to a surrounding volute casing.
Chatter	An irregular alternating motion of the parts of a relief valve due to the application of pressure where contact is made between the valve disk and the seat.
Coastdown	This occurs when the machine's driver is turned off and the suspect frequency is recorded as the speed decreases.
Common shaft	The individual shafts that exist in all machine-trains.
Displacement	The change in distance or position of an object relative to a reference point; the actual distance, off-centerline, of a rotating shaft as compared to a stationary reference, usually the machine housing.
Dynamic resonance	When the natural frequency of a rotating or dynamic structure, such as the rotor assembly in a fan, is energized, the rotating element will resonate.
Fast Fourier transform (FFT)	Converts a time-domain plot into its unique frequency components using a mathematical process.
First mode	The slightly eccentric rotation (off-center) of a shaft will generate a low-level frequency component that coincides with the rotating speed of the shaft.
Fourth mode	A shaft can flex or deform into mode shapes that will generate running-speed harmonics.
Frequency	The number of cycles completed by a periodic quantity in a unit time.
Frequency domain	A plane on which signal strength can be represented graphically as a function of frequency, instead of a function of time.
Fundamental or first critical speed	The lowest critical speed.
Fundamental train frequency	Generated by the precession of the cage as it rotates around the bearing races.

Gear mesh	Frequency is equal to the number of gear teeth times the running speed of the shaft.
Gravity	The gravitational attraction at the surface of a planet or other celestial body.
Harmonic motion	A periodic motion that is a sinusoidal function of time, that is, motion along a line given by equation $x = a \cos(kt + 0)$, where t is the time parameter, and a, k, and 0 are constants.
Harmonics	A sinusoidal component of a periodic wave, having a frequency that is an integral multiple of the fundamental frequency. Also known as harmonic component.
Helical gears	Gear wheels running on parallel axes, with teeth twisted oblique to the gear axis.
Herringbone gears	The equivalent of two helical gears of opposite hand placed side by side.
Hertz	Unit of frequency; a periodic oscillation has a frequency of n hertz if in 1 second it goes through n cycles.
Hydrodynamic	The study of the motion of a fluid and of the interactions of the fluid with its boundaries, especially in the incompressible inviscid case.
Imbalance	Any change in the state of equilibrium.
Laminar	Arranged in thin layers. Pertaining to viscous streamline flow without turbulence.
Low frequency cutoff	A frequency below which the gain of a system or device decreases rapidly.
Machine-train	A total machine including the driver, drive train, and machine.
Narrowband	The total energy within a user-selected range, or windows. Referring to a bandwith of 300 hertz or less.
Narrowband trending	Monitors the total energy for a specific bandwidth of vibration frequencies.
Natural frequency	The natural frequency of free vibration of a system. The frequency at which an undamped system with a single degree of freedom will oscillate upon momentary displacement from its rest position.

Node points	Where the shaft flexes into a double bend that crosses its true centerline.
Oil whip	Occurs when the clearance between the rotating shaft and sleeve bearing is allowed to close to a point approaching actual metal-to-metal contact. *See also* Oil whirl.
Oil whirl	An unstable free vibration whereby a fluid-film bearing has insufficient unit loading. The shaft centerline dynamic motion is usually circular in the direction of rotation. Oil whirl occurs at the oil flow velocity within the bearing, usually 40 to 49% of the shaft speed. Oil whip occurs when the whirl frequency coincides with (and becomes locked to) a shaft resonant frequency. (Oil whirl and whip can occur in any case where a fluid is between two cylindrical surfaces.)
Peak-to-peak	Amplitude of an alternating quantity measured from positive peak to negative peak.
Rate	The amount of change of some quantity during a time interval divided by the length of the time interval.
Rathbone chart	Provides levels of vibration severity that range from extremely smooth, best possible operating condition, to an absolute fault limit, or the maximum level where a machine can operate.
Resonance	A vibration of large amplitude in a mechanical system caused by a small periodic stimulus of the same or nearly the same period as the natural vibration period of the system.
Ringing	Method for exciting natural frequencies is to strike or excite a machine or structure with a timber or hammer.
Root-cause failure	Based on machine-train operation and how its dynamics affect the vibration spectrum.
Rotational frequencies	Related to the motion of the rolling elements, cage, and rings or races.
Running speed	The true rotational speed of the shaft or shafts.
Run-up	The machine's driver is turned on and records the amplitude and phase as the machine accelerates from dead-stop to full speed.

Sawtooth (waveform)	A waveform characterized by a slow rise time and a sharp fall, resembling a tooth of a saw.
Second mode	As the shaft rotates, the double bend shape will create two high spots as it passes the vibration transducer.
Signature	The characteristic pattern of target as displayed by detection and classification equipment.
Signature (FFT)	Usually applied to the vibration frequency spectrum unique to a particular machine or machine component, system, or subsystem at a specific location and point of time.
Static	A hissing, crackling, or other sudden sharp sound that tends to interfere with the reception, utilization, or enjoyment of desired signals or sounds. Without motion or change.
Static resonance	If the natural frequency of a stationary or non-dynamic structure, such as a casing, bearing pedestal, piping or other structure, is energized, the structure will resonate.
Synchronous	In step or in phase, as applied to two or more circuits, devices, or machines.
Third mode	A shaft can flex or deform into mode shapes that will generate running-speed harmonics.
Velocity	The time rate of change of position of a body; it is a vector quantity having direction as well as magnitude.
Vibration	A continuing periodic change in a displacement with respect to a fixed reference. In its general sense is a periodic motion. The motion will repeat itself in all its particulars after a certain interval of time.
Worm gears	Used for nonintersecting shafts at 90 degrees.
X-axis	A horizontal axis in a system of rectangular coordinates.
Y-axis	A vertical axis in a system of rectangular coordinates.

LIST OF ABBREVIATIONS

β	Contact angle (for roller = 0)
API	American Petroleum Institute
BD	Ball or roller diameter
BPFI	Ball-pass inner-race
cfm	gauge pressure
cpm	cycles per minute
cps	cycles per second
CSI	Computational Systems, Inc.
FFT	fast Fourier transform
F_{MAX}	maximum frequency
F_{MIN}	minimum frequency
f_r	relative speed between the inner- and outer-race (rps)
FTF	fundamental train frequency
Hz	hertz
in./sec	inches per second
IPS-PK	inches per second peak
kHz	kilohertz
MHz	megahertz
n	Number of balls or rollers
OEM	original equipment manufacturer

PD	pitch diameter
psig	cubic feet per minute
RMS	root mean square
rpm	revolutions per minute
VPM	vibrations per minute

INDEX

Printed and bound by CPI Group (UK) Ltd, Croydon, CR0 4YY

08/05/2025

01864813-0001